Contents

ebook

To access the ebook version of this Revision Guide visit
www.collins.co.uk/ebooks
and follow the step-by-step instructions.

Check your answers to the Practice Papers online:

www.leckieandleckie.co.uk

Higher Complete Revision and Practice

Complete Revision and Practice

This two-in-one Complete Revision and Practice book is designed to support you as students of Higher Human Biology. It can be used either in the classroom, for regular study and homework, or for exam revision. By combining a revision guide and two full sets of practice papers, this book includes everything you need to be fully familiar with the Higher Human Biology exam. As well as including all the core course content with practice opportunities, there is also some helpful advice on tackling assignment and exam preparation, links to the syllabus to help with targeted topic practice and easy reference with an index and Area Revision Test answers.

How to use the revision guide

The revision guide focuses on the final examination, which requires an in-depth knowledge of the course content to achieve a good grade. It is a revision aid which will also help you as the course continues. The guide follows the area structure of the course as detailed in the SQA guidelines to help you learn in a logical and user-friendly way.

Each area ends with a Revision Test giving you an opportunity to review how well you know the area content. Answers to these tests can be found at the end of the revision guide on page 124.

Throughout the revision guide, examples are given, where possible, to illustrate biological concepts. Use of examples is an important element for success in extended response questions.

Many assessment questions test your knowledge of the meaning of a particular word or phrase. To help you learn words and meanings, a glossary of terms has been included on page 116. One revision strategy you may consider is creating flashcards of these glossary terms and there are several free online programs which will help you make flashcards.

Use this revision guide as a starting point. Come back to it for the essential knowledge and skills that you need to tackle questions and understand concepts and processes in biology. This revision guide may also provide useful support during human biology lessons, and can be used as a quick reference in addition to other resources.

LECKIE
the education publisher for Scotland

Higher
HUMAN
BIOLOGY

For SQA 2019 and beyond

Revision + Practice
2 Books in 1

© 2020 Leckie

001/08012020

10 9 8 7 6 5 4 3 2

ISBN 9780008365295

Published by
Leckie & Leckie Ltd
An imprint of HarperCollinsPublishers
Westerhill Road, Bishopbriggs, Glasgow,
G64 2QT
T: 0844 576 8126 F: 0844 576 8131
leckiescotland@harpercollins.co.uk
www.leckiescotland.co.uk
1st Floor, Watermarque Building, Ringsend Road
Dublin 4, Ireland

Publisher: Sarah Mitchell
Project Manager: Lauren Murray

Special thanks to
Sarah Duxbury (cover design)
Jess White (proofread)
Jouve (layout and illustration)

Printed and bound in the UK using 100% Renewable Electricity at CPI Group (UK) Ltd

A CIP Catalogue record for this book is available from the British Library.

MIX
Paper from
responsible sources
FSC™ C007454

How to use the practice papers

This book contains two practice papers, which mirror the actual SQA exam as closely as possible in question style, level and layout. It is the perfect way to familiarise yourself with what the exam papers you will sit will look like.

The answer section, which can be found online at www.leckieandleckie.co.uk, contains worked answers to the practice papers, letting you know exactly where the marks are gained in an answer and how the right answer is arrived at. It also includes practical tips on how to tackle certain types of questions, details of how marks are awarded and advice on just what the examiners will be looking for.

The practice papers can be used in one of two ways:

1. You can complete an entire practice paper under exam conditions and mark it using the answer section. If you complete a practice paper this way it is important to make a list of all the key areas that you are having difficulty with by referring to the links to the syllabus on page 146, and concentrating your study time on these areas before attempting the next practice paper.
2. You can use the links to the syllabus to tackle individual topics. This allows you to focus on specific key parts of the course that you particularly want to revise. It also allows you to revise exam-style questions for just the topics you have covered.

Higher Human Biology

The Higher Human Biology course aims to develop your interest and enthusiasm for biology as it applies to human beings, as well as developing the skills required to function as a scientist in a modern-day context. The course is wide-ranging and draws on a number of different scientific disciplines to enable you to acquire a deeper understanding of cellular processes, physiological mechanisms and their impact on health, aspects of the nervous system and defence mechanisms as they apply to the human species. The skills developed will allow you to adapt to new situations, solve problems, make decisions based on evidence and evaluate how current developments in science might impact on your own health.

Structure
The Higher Human Biology course is made up of three areas:
*   Human cells (eight topics)
*   Physiology and health (eight topics)
*   Neurobiology and immunology (eight topics)

Course assessment

The course assessment has two elements:

National examination

The national examination gives you the opportunity to apply skills in both problem-solving and knowledge and understanding, answering questions from across the three areas of the course.

The exam carries a total of 120 marks, represents 80% of the overall course assessment marks and is made up of two sections:

- Paper 1 (25 marks): comprising 25 multiple choice questions lasting 40 minutes.
- Paper 2 (95 marks): made up of a mixture of restricted and extended response questions; there will be two or three extended response questions (totalling around 9–15 marks), one large question based on an experiment (worth around 5–9 marks) and one large data handling question (worth around 5–9 marks) lasting 2 hours 20 minutes.

You should be aware that techniques such as using gel electrophoresis, using a respirometer, measuring pulse rate, body mass index and blood pressure as well as a knowledge of the apparatus used such as a colorimeter and sphygmomanometer can be assessed here as well.

The assignment

The assignment assesses skills such as handling and processing data gathered through experimental work and research skills. It is worth 20 marks which are scaled up to 30 marks and represents 20% of the overall marks for the course assessment. It is recommended that no more than 8 hours are spent on the assignment with a maximum of 2 hours on the report stage.

There are two stages to the assignment:

1. Research stage:

Under some degree of supervision and control you will:

(i) plan, design and carry out an experiment in class that is relevant to your chosen topic in order to test an hypothesis and collect data

(ii) research the underlying biology of your chosen topic and find another source of data relevant to your experiment from an internet/literature source.

2. Report stage:

Here you will write a structured report of your findings under examination conditions. Your report will include relevant underlying biology, experimental data, analysis and comparison of your data with data from your other internet/literature source. Your report will contain an evaluation and allow a valid conclusion that relates to the aim to be drawn.

LECKIE
the education publisher
for Scotland

Higher
HUMAN
BIOLOGY

For SQA 2019 and beyond

Revision Guide

John Di Mambro, Deirdre McCarthy and Stuart White

Division and differentiation in human cells

Division of somatic and germline cells

A somatic cell is any body cell other than a cell involved in reproduction. Germline cells are haploid gametes, sperm and ova, and also the stem cells that divide to form gametes. A stem cell is an unspecialised somatic cell which can divide to make copies of itself (self-renew) and/or differentiate into specialised cells. Somatic cells divide by mitosis to form more somatic cells. Germline stem cells divide by mitosis to produce more germline cells and by meiosis to produce gametes. Mitosis preserves the diploid number of chromosomes, which is 46, arranged as 23 pairs of homologous chromosomes. When the germline cell divides by meiosis, it undergoes two divisions which separate homologous chromosomes and eventually separate the chromatids to form sperm and ova, which contain 23 single chromosomes.

TOP TIP

Mutations in germline cells are passed to offspring. Mutations in somatic cells are not passed to offspring.

Cellular differentiation

TOP TIP

Differentiated cells have a specific function and are unable to become any other type of cell.

Cellular differentiation is the process by which an unspecialised cell changes in order to carry out a specific function. A cell becomes a more specialised type of cell through gene expression. It develops more specialised functions by expressing the genes characteristic for that type of cell. For example, a stem cell can differentiate into a skin cell by switching on those genes which code for particular proteins, and by switching off the others which don't.

Stem cells

Stem cells are unspecialised somatic cells in animals that can divide to make copies of themselves (self-renew) and/or differentiate into specialised cells. They become specialised after differentiation when specific genes are expressed ('switched on') while others are not ('switched off'). This helps a wide variety of cells within animals to be differentiated from unspecialised stem cells, as shown in **Figure 1.1**.

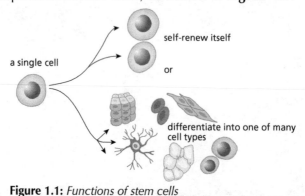

a single cell

self-renew itself

or

differentiate into one of many cell types

Figure 1.1: *Functions of stem cells*

Tissue stem cells

Tissue stem cells are involved in the growth, repair and renewal of the cells found in that tissue. For example, blood stem cells, found in the bone marrow, are able to form many components of blood such as red cells, lymphocytes, phagocytes and platelets. They are **multipotent** and capable of developing into a limited range of cell types.

Embryonic stem cells

The cells in **embryos** are **pluripotent** and can make all of the differentiated cell types of the body. All the genes in an embryonic stem cell can be switched on so these cells have the potential to differentiate into any type of cell. Under the right conditions in a laboratory, stem cells from the embryo can self-renew to form **embryonic stem cells**. Embryonic stem cells can offer effective treatments for injury and disease.

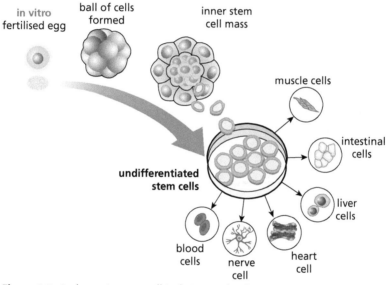

Figure 1.2: *Embryonic stem cell isolation and culture*

Research and therapeutic uses of stem cells

Stem cells are used in research:

- as **model cells** to observe how diseases develop
- for **drug testing** of potential medications.

Indeed, stem cell research provides valuable information on how gene regulation, differentiation and cell growth work.

Stem cells are used therapeutically:

- to repair damaged corneas
- to repair and regenerate damaged skin following a burn.

The ethical issues of stem cell use and the regulation of their use

The use of embryonic stem cells, which can offer effective treatments for diseases and injury, raises **ethical issues** in that the stem cells come from embryos that are destroyed.

Regulations ensure that the use of stem cells in research and therapy is carried out in accordance with UK law. This guarantees that procedures regarding the procurement of stem cells are conducted safely.

Cancer cells

Cancer cells divide excessively and produce a mass of abnormal cells called a **tumour**. They do not respond to regulatory signals and may fail to attach to one another. When cancer cells fail to attach to each other they spread through the body to form **secondary tumours**.

Structure and replication of DNA

Structure of DNA

Deoxyribonucleic acid or DNA is the substance that makes up the genetic material in cells and gives the cell its genotype.

DNA is composed of very long molecules made up of repeating chemical units called nucleotides.

A nucleotide has three chemical parts:

- An organic base
- A phosphate group
- A deoxyribose sugar

There are four different organic bases:

- Adenine (A)
- Guanine (G)
- Thymine (T)
- Cytosine (C)

TOP TIP

The sequence of bases in the DNA molecule forms the genetic code.

Nucleotides are linked together to make a long strand of DNA. This arrangement produces a sugar–phosphate backbone with the bases attached, as shown in **Figure 1.3**. The 5′ end has a phosphate attached to the fifth carbon of the deoxyribose sugar. The 3′ end has the third carbon of the deoxyribose sugar exposed.

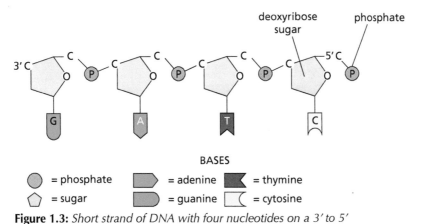

BASES

○ = phosphate   ⬠ = adenine   ◀ = thymine
⬠ = sugar   ⬛ = guanine   ◁ = cytosine

Figure 1.3: *Short strand of DNA with four nucleotides on a 3′ to 5′ sugar–phosphate backbone*

Two strands are joined together through **hydrogen bonds** that link **complementary** pairs of nucleotide bases.

Adenine always pairs with thymine. Guanine always pairs with cytosine.

The strands run in opposite directions from carbon 3′ at one end to carbon 5′ at the other end and are thus termed **antiparallel**, as shown in **Figure 1.4**.

The molecule is wound into a double-stranded helix, as shown in **Figure 1.5**.

Chromosomes consist of tightly coiled DNA and are packaged with associated proteins, as shown in **Figure 1.6**.

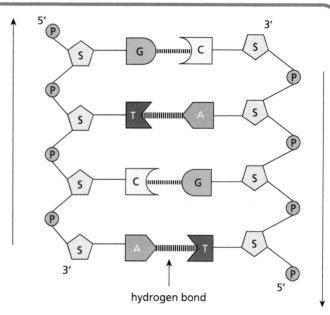

Figure 1.4: *Antiparallel strands of DNA*

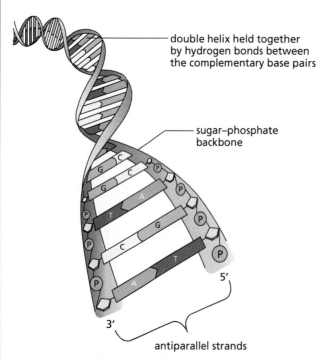

Figure 1.5: *Double helix of DNA showing antiparallel sugar–phosphate backbones joined by complementary base pairs*

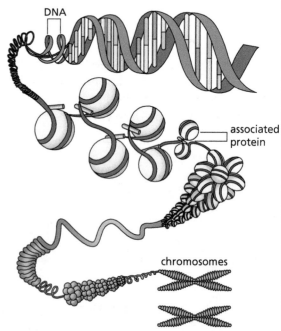

Figure 1.6: *Chromosome showing associated proteins*

Replication of DNA

It is essential that DNA is able to **replicate** (make more exact copies of itself) so that the newly formed cells have the exact same chromosome complement as the original cell.

In order for DNA replication to occur, certain substances must be present within the nucleus:

- The DNA template
- Free DNA nucleotide bases (all four types)
- Enzymes (such as **DNA polymerase** and **ligase**)
- **Primers** (their presence is required to begin replication)
- **ATP** (to supply energy for the process)

> ### TOP TIP
> A primer is a short sequence of nucleotides which binds to the 3' end of the template DNA strand allowing polymerase to bind.

DNA replication is shown in **Figure 1.7** and involves the following stages:

1. DNA is unwound and hydrogen bonds between bases are broken to form two templates.

2. Primers bind to the 3' end of the template DNA strand.

3. DNA polymerase binds to the primers and adds DNA nucleotides, using complementary base pairing, to the deoxyribose 3' end of the new DNA strand that is forming.

4. As the DNA polymerase enzyme only adds free nucleotides in one direction then one strand is replicated continuously - the leading strand and the other is replicated in fragments.

> ### TOP TIP
> DNA polymerase is only able to add DNA nucleotides to the 3' end of the DNA template, resulting in a leading strand and a lagging strand.

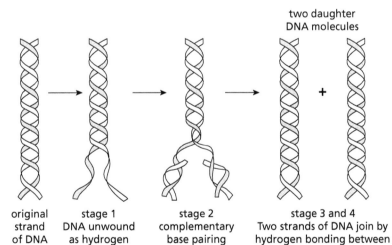

two daughter DNA molecules

| original strand of DNA | stage 1 DNA unwound as hydrogen bonds break | stage 2 complementary base pairing | stage 3 and 4 Two strands of DNA join by hydrogen bonding between complementary bases. They twist to form a double helix. |

Figure 1.7: *Summary of stages of DNA replication*

Leading strand

On the leading strand, DNA polymerase adds DNA nucleotides using complementary base pairing to the deoxyribose (3') end of the newly forming DNA strand in a continuous fashion.

Lagging strand

The lagging strand is antiparallel to the leading strand but replication cannot begin at the 5' end. Instead it is replicated in fragments, starting with primer at the 3' end. Nucleotides are added in a discontinuous fashion as more primer is added to the lagging strand, as shown in **Figure 1.8**.

The third stage involves the use of the enzyme ligase, which joins together the small fragments of newly formed DNA in the lagging strand to form a sugar–phosphate backbone.

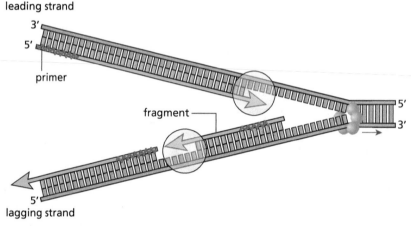

Figure 1.8: *Leading and lagging strand replication*

TOP TIP

To produce new cells and pass on the correct genetic instructions to new generations, DNA replication must create identical copies of the information contained in the DNA.

TOP TIP

The exact sequence of the bases makes one species distinct from another.

Amplification and detection of DNA sequences

The **polymerase chain reaction** (PCR) is a technique for the **amplification** of DNA in vitro and involves the use of a **thermal cycler**.

In PCR the complementary primers are chosen to target specific sequences at the two ends of the region of DNA being amplified.

Similar to DNA replication, PCR has some initial requirements for the process to occur. These are:

- DNA template.
- Free DNA nucleotides (all four types).
- Heat-tolerant DNA polymerase (enzyme).
- Primers – artificially made, short, single strands of DNA that use bases complementary to those at either end of the target DNA fragment to be copied.

Figure 1.9 shows the general stages of the PCR cycle.

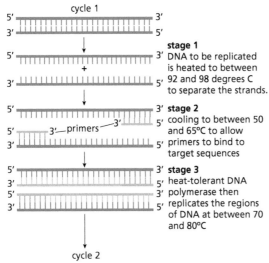

Figure 1.9: *The stages of PCR*

> **TOP TIP**
>
> In vitro literally means 'in glass' but is used for procedures carried out in a lab context.

> **TOP TIP**
>
> When a thermal cycler is used, the PCR cycle can be repeated many times.

1. DNA heated to between 92 and 98°C to separate strands, then cooled to between 50 and 65°C for primer binding of complementary primers.

2. Cooling allows primers to bind to target sequences.

3. Heat-tolerant DNA polymerase then replicates the region of DNA at between 70 and 80°C.

Repeated cycles of heating and cooling amplify this region of DNA, as shown in **Figure 1.10**.

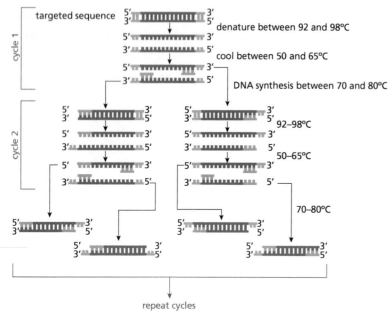

Figure 1.10: *Repeated cycles of PCR amplifying the original DNA strand*

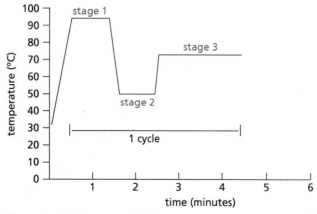

Figure 1.11: *The stages of PCR in a time vs temperature graph*

When a thermal cycler is used, the PCR cycle can be repeated many times.

Practical applications of PCR

The DNA amplified by PCR can be used in many ways, such as:

- to help solve crimes
- to settle paternity suits
- to diagnose genetic disorders.

TOP TIP

PCR is useful in solving crimes and determining paternity.

Gene expression

Gene expression

Gene expression (**Figure 1.12**) involves the transcription and translation of DNA sequences. The genetic code processed by transcription and translation is found in all forms of life.

An organism's **phenotype** is determined by the proteins produced as a result of gene expression. The genome of an organism is its entire hereditary information encoded in DNA. Not all the genes are expressed in every cell in an organism.

TOP TIP

Only a fraction of the genes in a cell are expressed.

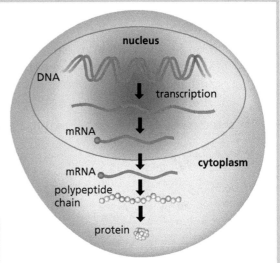

Figure 1.12: *Overview of gene expression*

Ribonucleic acid (RNA)

RNA is vital to the process of protein synthesis. It has some similarities to DNA but differs in three ways as shown in **Figure 1.13**:

1. RNA is a single stranded molecule (DNA is double stranded).

2. RNA has the same bases as DNA except for uracil, which replaces thymine.

3. RNA has the sugar ribose whereas DNA has the sugar deoxyribose.

Feature	DNA	RNA
number of strands	2	1
bases	A T G C	A U G C
sugar	deoxyribose	ribose

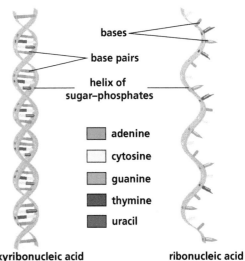

Figure 1.13: *DNA and RNA*

The main forms of RNA are:

1. Messenger RNA (mRNA) which carries a copy of the DNA code from the nucleus to the ribosome. It is a single stranded molecule and triplets of bases along this strand are known as **codons**.

2. Ribosomal RNA (rRNA), which, along with ribosomal protein, forms protein-synthesising structures called **ribosomes**.

3. Transfer RNA (tRNA) molecules each carry a specific amino acid to the ribosome and are involved in the second part of protein synthesis. They have a folded shape and an **anticodon** at one end and an attachment site for a specific amino acid at the other end.

The three types of RNA are shown in **Figure 1.14**.

TOP TIP

The sequence of bases which form an anticodon are complementary to a specific codon on the mRNA.

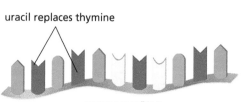
uracil replaces thymine
messenger RNA

ribosome
ribosomal RNA

amino acid
transfer RNA

Figure 1.14: *Types of RNA*

Transcription

Transcription, shown in **Figure 1.15**, occurs in the nucleus of a cell and is the first phase of gene expression. The particular DNA sequence for the gene to be expressed is transcribed into a primary transcript of mRNA in the following process:

1. **RNA polymerase** moves along the DNA and unwinds the double helix.

2. The polymerase breaks the hydrogen bonds between the base pairs.

3. As RNA polymerase breaks the bonds, it synthesises a primary transcript of mRNA on the DNA template strand using free RNA nucleotides. These RNA nucleotides form hydrogen bonds with the exposed DNA bases by complementary base pairing.

4. A primary RNA transcript is formed.

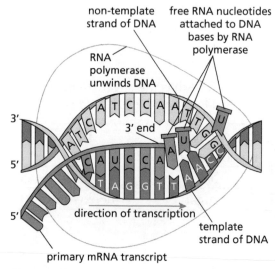
non-template strand of DNA
free RNA nucleotides attached to DNA bases by RNA polymerase
RNA polymerase unwinds DNA
3′ end
direction of transcription
template strand of DNA
primary mRNA transcript

Figure 1.15: *Transcription*

RNA splicing

The primary transcript is made of **introns** and **exons**.

5. The introns of the primary transcript of mRNA are **non-coding** and are removed.

6. The exons are **coding** regions and are joined together to form a **mature mRNA transcript**.

7. This process is called **RNA splicing**.

This process is summarised in **Figure 1.16**.

After leaving the nucleus through a pore in the nuclear membrane, the mature transcript travels through the cytoplasm to a ribosome for the next stage of gene expression.

TOP TIP

The order of exons is not changed during splicing.

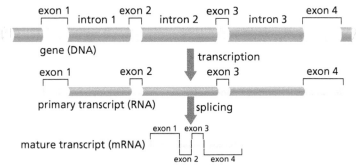

Figure 1.16: *RNA splicing from primary to mature transcript*

Translation

The mature transcript of mRNA arrives and attaches itself to a site on the ribosome. The sequence of codons on the mRNA is read as the complementary tRNA anticodons carrying the specific amino acids are brought to the ribosome.

As the ribosome moves along, the mRNA molecule is read and amino acids bind together using peptide bonds. tRNA then exits from the ribosome. As the chain grows, the sequence of amino acids becomes a **polypeptide**.

The starting and end points of the chain formed are determined by triplets of mRNA called start and **stop codons**, respectively. The mRNA codon AUG, which codes for the amino acid methionine, also acts as a **start codon**. The mRNA codons UAA, UGA and UAG do not code for a specific amino acid. Instead they act as stop codons, which terminate the polypeptide chain formation.

Translation takes place in the following stages, as shown in **Figure 1.17**:

1. mRNA attaches to a site on the ribosome.
2. The first codon of the mRNA is called a start codon and this starts the translation process.
3. Codons bond to complementary bases on the anticodon of tRNA, which transports specific amino acids to the ribosome. The anticodon on a tRNA molecule is specific to only one amino acid.
4. The amino acids form a chain, joined together by peptide bonds.
5. This chain is called a polypeptide.
6. The last codon of the mRNA is called a stop codon and this stops the translation process.
7. The tRNA leaves the ribosome as the specific amino acid it was carrying is joined to the previous one forming a polypeptide chain.
8. The way in which polypeptide chains are assembled determines the structure and function of the finished protein.

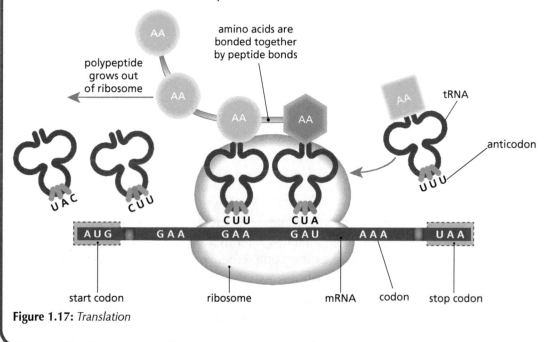

Figure 1.17: *Translation*

Proteins

Different proteins can be expressed from one gene as a result of alternative RNA splicing.

Different mature mRNA molecules are produced from the same primary transcript. These mature mRNA transcripts contain different exons.

Some may have phosphate or carbohydrate added while others may be cut or combined to form new polypeptide chains, as shown in **Figure 1.18**.

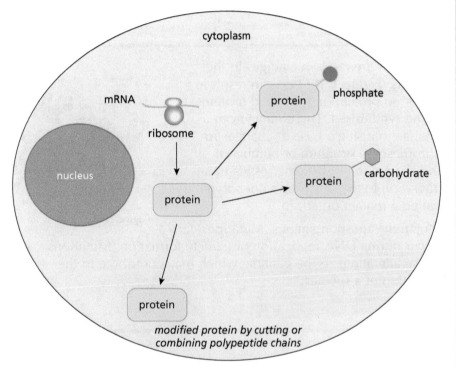

Figure 1.18: *Potential modifications of translated protein*

> **TOP TIP**
>
> Proteins are usually folded to form a 3D shape held together by hydrogen bonds and other interactions.

Protein structure

Proteins produced after translation have a variety of structures which determine their functions. The sequence of amino acids in a polypeptide chain determines the structure and function of the protein.

There are 20 different types of amino acids which make up proteins.

Protein shape variation arises from folding, creating three-dimensional shapes. Folding of polypeptide chains is caused by hydrogen bonds and interactions between amino acids.

Mutations

Mutations

Mutations are random changes in the genome (**Figure 1.19**) that can result in either no protein or an altered protein being synthesised. These range from a change in a single base to changes in chromosome structure or number. If the change to the genome results in an alteration to the organism's phenotype, it is called a mutant organism.

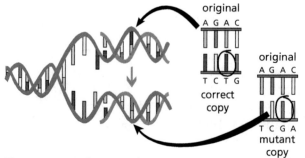

Figure 1.19: *A change in the genome*

Mutations are spontaneous. Mutations can occur during DNA replication or gamete formation. Mutations cause alterations to the genome which may contribute to the evolution of a species.

> **TOP TIP**
>
> Remember that phenotype is affected by environmental conditions as well as by the genotype.

Single gene mutations

A single nucleotide base change results in single gene mutations. The DNA nucleotide sequence is altered by the **substitution**, **insertion** or **deletion** of nucleotides. The impact on the protein synthesised can be either minor or major, depending on the type of mutation. The table below shows single gene mutations and these are illustrated in **Figure 1.20**.

Single gene mutation	DNA base sequence	Impact of gene mutation
none	ATG CGT CGA	
substitution	ATG CCT CGA	may result in a non-functional protein or have little effect
insertion	ATG CGG TCG	major change to protein structure
deletion	ATG CTC GA	major change to protein structure

If a single nucleotide is substituted then this will only change one codon, resulting in a minor change to the protein produced. Single nucleotide substitutions include **missense**, **nonsense** and **splice-site** mutations.

> **TOP TIP**
>
> When asked to describe the effect of a mutation on a gene you should mention the nucleotide base sequence or codons in your answer. When asked to describe the effect of a mutation on the structure of a protein you should mention the amino acid sequence in your answer.

Substitution	Change	End effect on protein produced
missense	one codon to another	different amino acid translated
		possible change in protein shape
		may produce a non-functional protein or have little effect on the protein
nonsense	a codon to a stop codon	shortens the protein
		may become non-functional or its function will be changed
splice-site	change in nucleotide at a splice-site (between intron and exon)	may prevent splicing
		this results in a very different protein being synthesised because introns may be left in the primary transcript
		some exons may not be included in the transcript

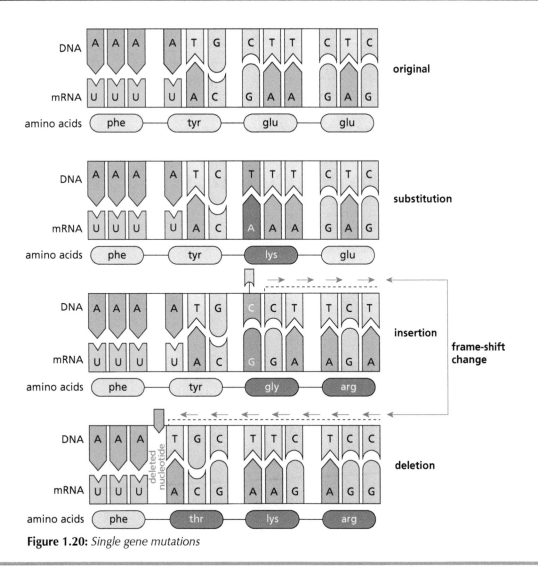

Figure 1.20: *Single gene mutations*

Nucleotide insertions or deletions result in **frame-shift mutations**. When the altered codon is translated at the ribosome into an amino acid, all the subsequent codons and thus amino acids are changed. This may have a major effect such as a faulty, non-functional or alternative protein. Mutations in some non-coding DNA sequences can result in changes to the way certain genes are expressed. The frame-shift effects of insertion and deletion are shown in **Figure 1.21**.

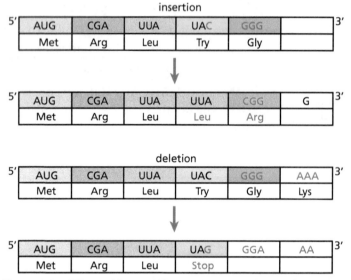

TOP TIP

Many human conditions are now known to be caused by mutations. These include sickle-cell disease (missense), Duchenne muscular dystrophy (nonsense) and cystic fibrosis (frame-shift deletion).

Figure 1.21: *Effects of frame-shift mutations on amino acid sequences*

Chromosome structure mutations

Chromosome structure mutations are those which affect whole chromosomes or sections of the genome. They are alterations to the structure of one or more chromosomes. They include:

- **duplication**
- **deletion**
- **inversion**
- **translocation**

Each of the chromosome mutations are described in the table below and shown in **Figure 1.22**.

Mutation	Description
duplication	produced when extra copies of genes on a homologous chromosome are generated
deletion	results when a section of chromosome is removed
inversion	a section of a chromosome is reversed
translocation	the piece of chromosome detaches from one chromosome and moves to a new position on a non-homologous chromosome

The mutated chromosome segments are often large and affect many genes. This results in large alterations in the genome and the proteins produced, many of which will be defective. Substantial changes in chromosome mutations can make them lethal.

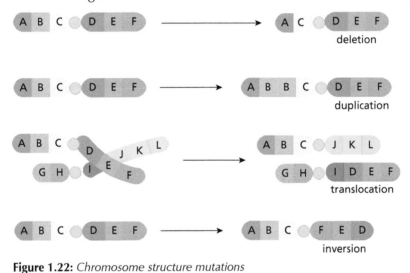

Figure 1.22: *Chromosome structure mutations*

Human genomics

Sequencing DNA

Genomic sequencing is a process in which the order of the nucleotide bases along an organism's genome is determined. A genome is made up of genes and other DNA sequences which do not code for proteins.

The sequence of nucleotide bases can be determined for individual genes and for entire genomes. This helps identify genes and genome sequences that are responsible for particular diseases and conditions, and helps identify genomic mutations and variations.

Bioinformatics is used to compare sequence data. Analysis particularly of the human genome is one of the main achievements of bioinformatics. Future developments include a fuller understanding of how the human genome works leading to enhanced use of targeted drugs and individualised therapy.

> **TOP TIP**
>
> Computer technology is used now to identify base sequences by looking for sequences similar to known genes.

Personalised medicine

Due to advances in technology, an individual human genome can now be sequenced cheaply and quickly, for a fraction of that cost, and in days rather than years. Therefore, obtaining an individual's personal genome is relatively easy and this information could be used in a number of ways in the future.

One way in which this information could be used is to identify mutations within the genome. As discussed before, these can either be harmful (the changes result in non-functional protein forming) or neutral (the changes have no negative effect). Analysis of an individual's genome can often predict the likelihood of certain diseases developing.

Analysis of an individual's genome leads to **personalised medicine**. Using **pharmacogenetics** the success of a particular treatment can be predicted. The identification of the genomic change responsible for a genetic disorder enables a specific treatment to be applied. Individual treatments are personalised, as shown in **Figure 1.23**, and more likely to succeed.

> **TOP TIP**
>
> The Human Genome Project, which was finished in 2003, sequenced three billion nucleotide bases and found around 20,000 genes in the human genome.

> **TOP TIP**
>
> Personalised medicine can be used in the selection of the best drugs and dosage for a patient's particular disease.

Several diseases and conditions are the result of a combination of genetic and environmental factors. This causes difficulties when it comes to treatment. These diseases can be complex and, additionally, agents such as viruses can be involved.

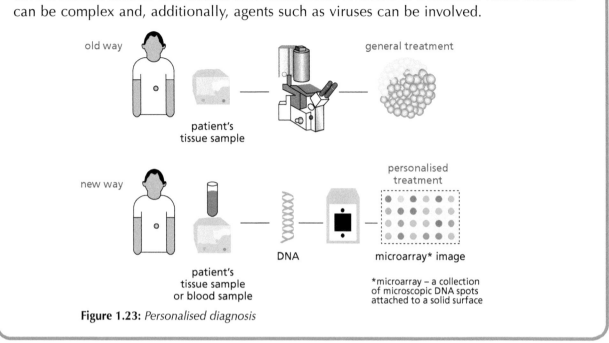

Figure 1.23: *Personalised diagnosis*

Metabolic pathways

Cell metabolism

Cell metabolism is the collective term for the thousands of biochemical reactions that occur within a living cell. A metabolic pathway is a series of chemical reactions occurring within a cell. Metabolism encompasses the integrated and controlled pathways of enzyme-catalysed reactions within a cell.

Anabolic and catabolic pathways

Anabolism

Anabolic pathways require energy and involve biosynthetic processes, as shown in **Figure 1.24**.

Catabolism

Catabolic pathways release energy and involve the breakdown of molecules, as shown in **Figure 1.25**.

These pathways can have reversible and irreversible stages and alternative routes. Metabolic pathways may exist that can bypass stages in a pathway, as shown in **Figure 1.26**.

An example of a linked pathway can be seen in **Figure 1.27**.

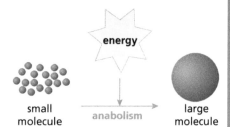

Figure 1.24: *A simple anabolic pathway*

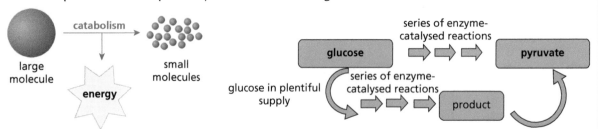

Figure 1.25: *A simple catabolic pathway*   **Figure 1.26:** *Bypass pathway when glucose is in plentiful supply*

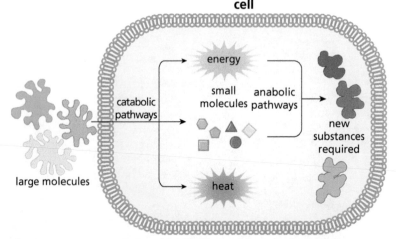

Figure 1.27: *How catabolic and anabolic reactions can be linked*

Control of metabolic pathways

Metabolic pathways can be controlled in a number of ways, for example, through the presence or absence of particular enzymes and the regulation of the rate of reaction of key enzymes within a pathway.

Induced fit

Induced fit describes the change when a molecule of substrate enters the active site. The active site changes shape making it fit very closely round the substrate molecule. This is shown in **Figure 1.28**.

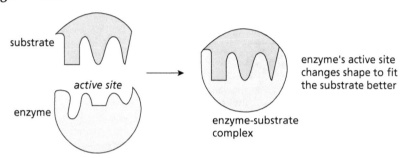

Figure 1.28: *Induced fit of enzyme's active site interacting with specific substrate*

The active site is not a rigid structure. It is flexible and dynamic. The enzyme is specific to its substrate. It will only bind to one particular substrate because the shape of the active site has to match the shape of the substrate.

The substrate has a high affinity for the active site. The resulting products have a low affinity for the active site and so leave the active site.

Activation energy is needed to break chemical bonds in the reactant chemicals. Induced fit lowers the activation energy needed by reactants and increases the reaction rate.

TOP TIP

The majority of metabolic reactions are reversible and the presence of a substrate or the removal of a product will drive a sequence of reactions in a particular direction.

Control of metabolic pathways

Control of metabolic pathways is achieved through **competitive**, **non-competitive** and **feedback inhibition** of enzymes.

- Molecules of a **competitive inhibitor** compete with molecules of the substrate for the active sites on the enzyme, as shown in **Figure 1.29**. The inhibitor is able to do this because its molecular structure is similar to that of the substrate and it can attach itself to the enzyme's active site. Competitive inhibition can be reversed by increasing substrate concentration. This is shown in **Figure 1.30**.

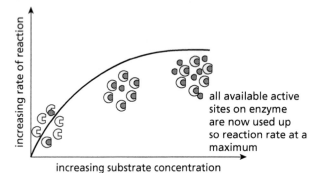

all available active sites on enzyme are now used up so reaction rate at a maximum

Figure 1.30: *Effect of increasing substrate concentration on the rate of reaction*

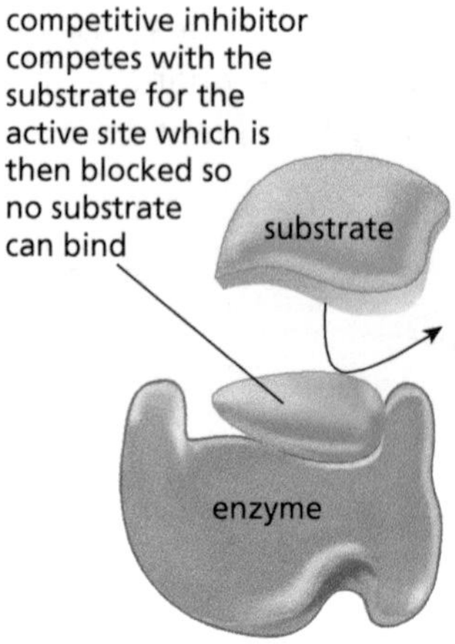

competitive inhibitor competes with the substrate for the active site which is then blocked so no substrate can bind

Figure 1.29: *Action of a competitive inhibitor*

- A non-competitive inhibitor does not combine directly with an enzyme's active site. Instead it becomes attached to a place other than the active site and changes the shape of the active site thus preventing the substrate from binding. This is shown in **Figure 1.31**.

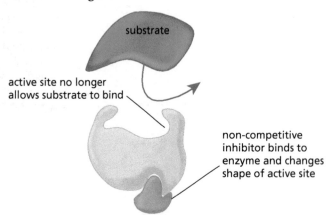

active site no longer allows substrate to bind

non-competitive inhibitor binds to enzyme and changes shape of active site

TOP TIP

Non-competitive inhibition cannot be reversed by increasing substrate concentration.

Figure 1.31: *Action of a non-competitive inhibitor*

- Control of metabolic pathways is also achieved through feedback inhibition of enzymes. The rate at which some metabolic pathways progress is controlled by a build-up of the end-product to a critical concentration. In feedback inhibition, the end-product binds to an earlier enzyme in the metabolic pathway. This alters the shape of this enzyme's active site and stops the pathway, as shown in **Figure 1.32**. This prevents too much end-product from being produced. As the concentration of the end-product drops, inhibition ceases and the pathway resumes again. This is shown in **Figure 1.33**.

> **TOP TIP**
>
> The presence of a substrate or the removal of a product can also control metabolic reactions by driving them in a particular direction.

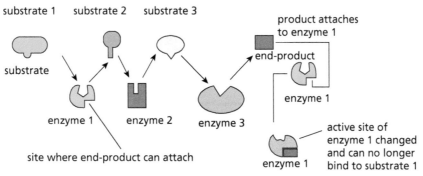

Figure 1.32: *Action of end-product causing change in shape of the active site of enzyme 1*

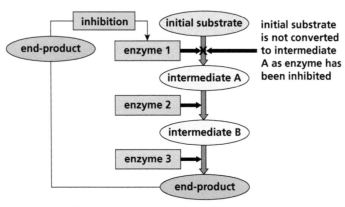

Figure 1.33: *Feedback inhibition of a metabolic pathway*

Cellular respiration

ATP

ATP is used to transfer energy to synthetic pathways and other cellular processes where energy is required.

Phosphorylation is an enzyme-controlled process by which a phosphate group is added to a molecule. For example, phosphorylation occurs when P_i combines with low-energy ADP to form high-energy ATP, as shown in **Figure 1.34**.

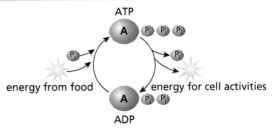

Figure 1.34: *Transfer of energy from ATP breakdown*

Metabolic pathways of cellular respiration

There are three stages in cellular respiration:
- **Glycolysis** (takes place in cytoplasm).
- **Citric acid cycle** (takes place in the **matrix of the mitochondria**).
- **Electron transport chain** (takes place on the inner membrane of the mitochondria).

During this process, glucose is broken down, and hydrogen ions and electrons are removed from substrates by **dehydrogenase enzymes**, releasing ATP.

Glycolysis

Glycolysis takes place in the cytoplasm. Glycolysis is the breakdown of glucose (6C) to **pyruvate** (3C).

The phosphorylation of glucose and intermediates in glycolysis in an **energy investment stage** leads to the generation of more ATP in an **energy pay-off stage** giving a **net gain** of ATP.

To start the process energy from ATP is used up but eventually more ATP is generated than was used up resulting in an overall net gain. This is shown in **Figure 1.35**.

Hydrogen ions and electrons released at this stage are transported by the coenzyme NAD to the electron transport chain.

In the absence of oxygen, pyruvate undergoes fermentation to lactate, as shown in **Figure 1.36**.

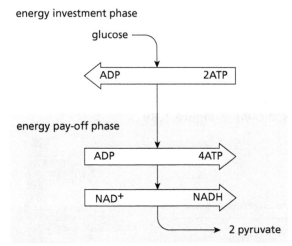

Figure 1.35: *The two phases of glycolysis*

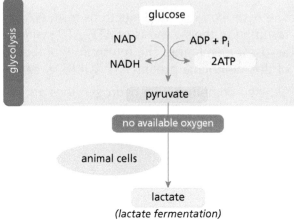

Figure 1.36: *Fermentation in the cytoplasm of animal cells*

Citric acid cycle

This stage takes place under aerobic conditions in the matrix of the mitochondria. Pyruvate is broken down to an acetyl group that combines with coenzyme A to form acetyl coenzyme A as shown in **Figure 1.37**.

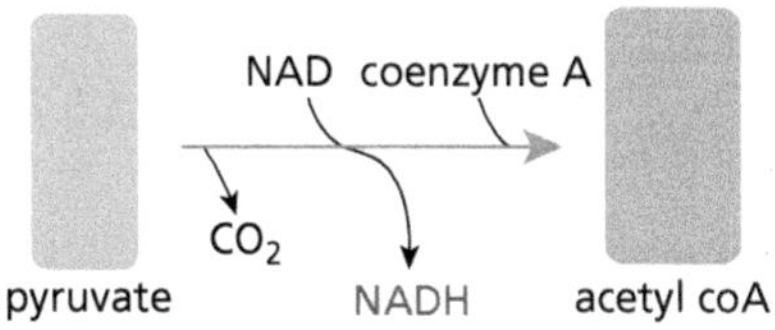

The acetyl group from the coenzyme A combines with oxaloacetate to form citrate. The enzyme-mediated stages of the cycle follow, resulting in the generation of ATP, the release of carbon dioxide and the regeneration of oxaloacetate. This is summarised in **Figure 1.38**.

Figure 1.37: *Formation of acetyl coenzyme A from coenzyme A (CoA)*

Dehydrogenase enzymes remove hydrogen ions and electrons, which are passed to the coenzyme NAD (forming NADH) in glycolysis and citric acid pathways. Two ATP are produced during the citric acid cycle.

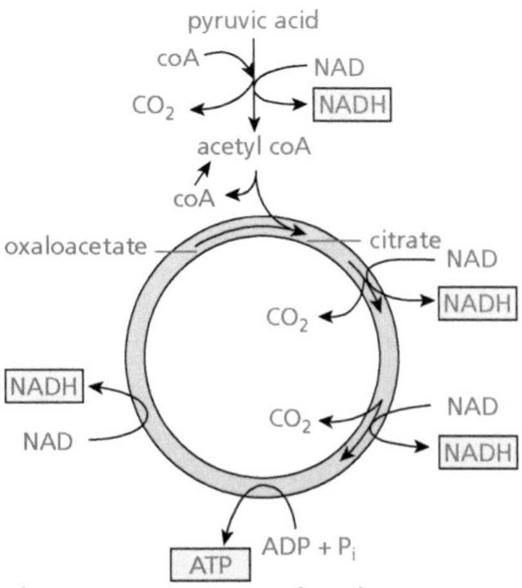

Figure 1.38: *Citric acid cycle*

> ***TOP TIP***
>
> Like many carrier molecules, coenzyme A can be recycled and used over and over again.

Electron transport chain

This stage takes place on the folded inner membrane of the mitochondrion. The electron transport chain is a series of carrier proteins attached to the inner mitochondrial membrane.

The hydrogen ions and electrons from NADH are passed to the electron transport chain, resulting in the synthesis of ATP. ATP synthesis involves electrons pumping hydrogen ions across a membrane. The return flow of these ions back through the membrane rotates part of the membrane protein **ATP synthase**, catalysing the synthesis of ATP.

Oxygen combines with hydrogen ions and electrons, forming water. Most ATP are produced by the electron transport chain. This is shown in **Figure 1.39**.

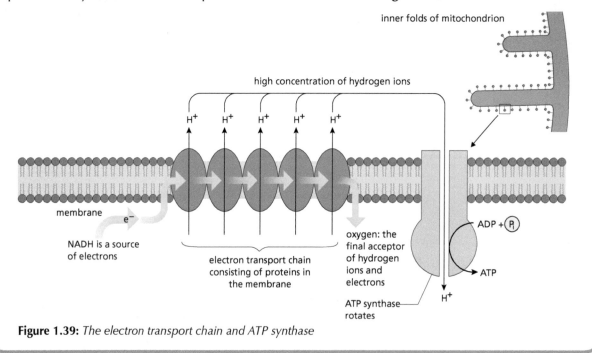

Figure 1.39: *The electron transport chain and ATP synthase*

Energy systems in muscle cells

Lactate metabolism

During vigorous exercise, the muscle cells do not get sufficient oxygen to support the electron transport chain. Under these conditions, pyruvate is converted to lactate. This conversion involves the transfer of hydrogen (from NADH produced in glycolysis) to pyruvate in order to produce lactate. This regenerates the NAD needed to maintain ATP production through glycolysis. Lactate accumulates in muscle causing fatigue. This process is shown in **Figure 1.40**.

The oxygen debt is repaid when exercise is complete, allowing respiration to provide the energy to convert lactate back to pyruvate and glucose in the liver.

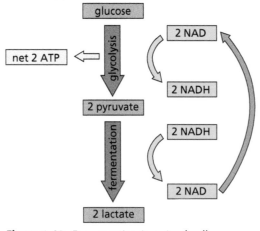

Figure 1.40: *Fermentation in animal cells*

Types of skeletal muscle fibres

Slow twitch (type 1) muscle fibres are good for endurance activities such as long-distance running. Slow twitch muscle fibres rely on aerobic respiration to generate ATP and have many mitochondria, a large blood supply and a high concentration of the oxygen-storing protein myoglobin. The major storage fuel of slow twitch muscle fibres is fats.

Fast twitch (type 2) muscle fibres are good for activities such as sprinting or weightlifting. Fast twitch muscle fibres can only generate ATP through glycolysis. They have few mitochondria and a lower blood supply than slow twitch muscle fibres. The major storage fuel of fast twitch muscle fibres is glycogen.

Slow twitch muscle fibres contract more slowly, but can sustain contractions for longer, and are therefore better for endurance activities. Fast twitch muscle fibres contract more quickly, over short periods, and are therefore better for activities which require short bursts of activity.

Most human muscle tissue contains a mixture of both slow twitch and fast twitch muscle fibres. Athletes exhibit distinct patterns of muscle fibres which reflect their sporting activities.

Feature	Type of skeletal muscle fibre	
	Slow twitch (type 1)	Fast twitch (type 2)
activity most suitable for	endurance activities	sprinting/weightlifting
method of generating ATP	aerobic respiration	glycolysis only
mitochondria number	many	few
blood supply	large	lower
myoglobin present	yes	no
major storage fuels	fats	glycogen
contraction speed	slow	fast
contraction duration	long	short

TOP TIP

Endurance activities include long-distance running, cycling or cross-country skiing.

Area 1 Revision Test

Key area 1 – Division and differentiation in human cells

1. (a) Draw lines to match each of the following terms with its correct description.   (3)

Term	Description

Term   **Description**

| Stem cell | A mass of abnormal cells |

| Tumour | Gamete (such as sperm or ovum) |

| Germline cell | Unspecialised somatic cell that can divide to make copies of itself (self-renew) and/or differentiate into specialised cells |

(b) State **one** current therapeutic use of stem cells.   (1)

Key area 2 – Structure and replication of DNA

2. The following diagram shows two strands of DNA.

(a) Name bond X.   (1)

(b) Name molecules Y and Z.   (2)

(c) Name the base which is complementary to cytosine.   (1)

(d) State the other **two** bases found in DNA.   (2)

(e) State the term used to describe the arrangement of the two DNA strands shown.   (1)

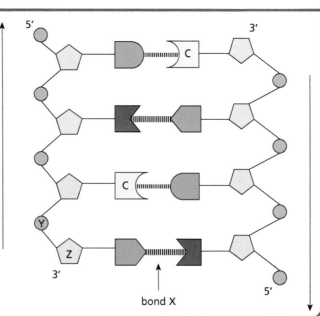

Key area 3 – Gene expression

3. Decide if each of the following statements about the structure of DNA and RNA is **True** or **False** and tick (✔) the correct box.

   If the answer is **False**, write the correct word(s) in the **Correction** box to replace the word underlined in the statement. (3)

Statement	True	False	Correction
RNA is a single stranded molecule.			
DNA contains the base uracil.			
An RNA nucleotide contains a base, a phosphate and a deoxyribose sugar.			

4. The diagram below shows a strand of RNA during protein synthesis.

   (a) Name X1 and Y1. (2)

   (b) State the name of process Z. (1)

   (c) State the location of transcription within the cell. (1)

   (d) Name the molecule found in the cytoplasm which carries a specific amino acid to a ribosome. (1)

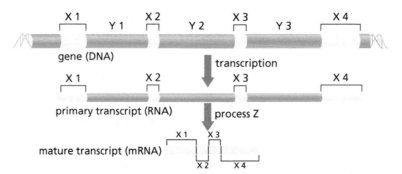

Key area 4 – Mutations

5.   An original DNA base sequence is shown below.

ATG CGT ACG

The table below shows three mutations.

Complete the table by inserting the names of the single gene mutations and if they cause a frame-shift mutation by inserting yes or no.   (3)

Single gene mutation	DNA base sequence	Frame-shift mutation (yes/no)
	ATG CGT CCG	
	ATC GCG TAC	
	ATG CTA CGC	

Key area 5 – Human genomics

6.   The graph below shows the stages of PCR with time (minutes) vs. temperature (°C).

(a) State the purpose of PCR.   (1)

(b) Explain the purpose of heating in stage 1.   (1)

(c) Describe the function of the primers used in stage 2.   (1)

(d) Name the enzyme used in stage 3.   (1)

(e) If one cycle takes 4·5 minutes, calculate the number of molecules of DNA that were produced after 45 minutes from one original DNA strand.   (1)

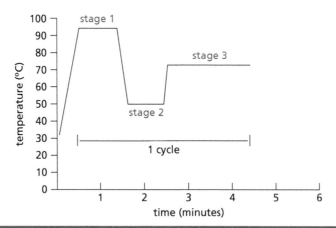

Key area 6 – Metabolic pathways

7. The diagram below shows a metabolic pathway.

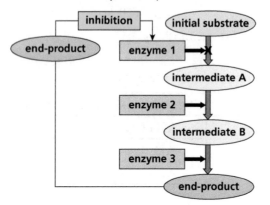

(a) This metabolic pathway requires energy. State the term used for this kind of pathway. (1)

(b) The end-product inhibits enzyme 1 by attaching to its active site. State the term given to this type of inhibition. (1)

(c) Predict the effect of increasing the concentration of the end-product on the concentration of intermediate B. (1)

(d) State the term given to this kind of inhibition. (1)

Key area 7 – Cellular respiration

8. The diagram below shows the three stages of cellular respiration.

   (a) Name stage X. (1)

   (b) Name molecule Z. (1)

   (c) Describe the effect of a high concentration of ATP or citrate being produced on the rate of cellular respiration. (1)

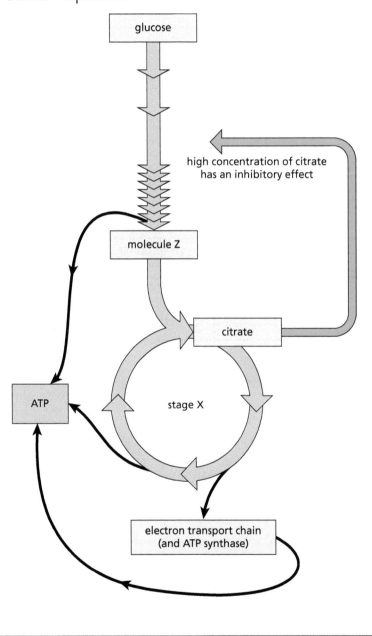

Key area 8 – Energy systems in muscle cells

9. The table below states some of the features of slow twitch and fast twitch skeletal muscle fibres. Complete the table by filling in the appropriate words in each empty box. (5)

Feature	Type of skeletal muscle fibre	
	Slow twitch (type 1)	Fast twitch (type 2)
activity most suitable for		
myoglobin present		
major storage fuels		
contraction speed		
contraction duration		

10. An experiment was carried out to investigate the effect of different glucose concentrations on the rate of respiration in yeast. Methylene blue will change from blue to colourless as the yeast respires. The time taken will depend upon the rate of respiration in the yeast.

The results are shown below.

Glucose concentration (%)	Time taken for methylene blue to change from blue to colourless (seconds)
1	90
2	64
3	59
4	52
5	46
6	38
7	30

(a) State a conclusion that can be made from these results. (1)

(b) Calculate the percentage decrease for time taken for methylene blue to change to colourless between glucose concentrations of 1% and 7%. (1)

(c) Predict the time taken for methylene blue to change to colourless if the glucose concentration was increased to 8%. (1)

(d) Between which two glucose concentrations did the time taken for methylene blue to change to colourless decrease by the least? (1)

Gamete production and fertilisation

Gamete production in the testes

Male reproductive system

Male gametes are produced in the testes (**Figure 2.1**) from germline cells by meiotic division, which reduces the chromosome number from diploid to haploid.

Within the testes are the coiled **seminiferous tubules** where sperm production takes place (**Figure 2.2**). Between the seminiferous tubules are **interstitial cells** which produce the male hormone **testosterone**, which is transported in the bloodstream. In addition to stimulating sperm production,

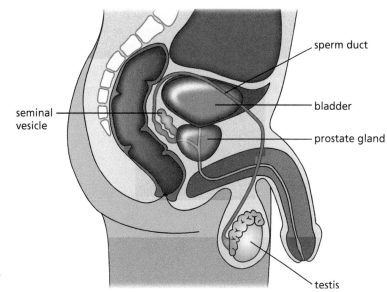

Figure 2.1: *Structure of male reproductive system*

testosterone also promotes the development of secondary sexual characteristics such as deepening voice and facial hair. Testosterone acts on the **prostate gland** and the **seminal vesicles**. The action of these two organs produces fluid which, when added to the sperm, is collectively called **semen**. These fluids maintain the mobility and viability of sperm.

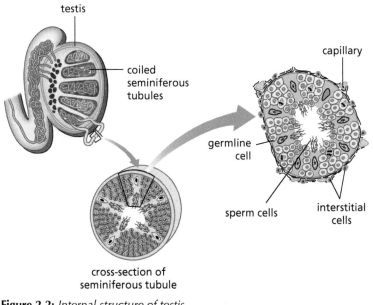

Figure 2.2: *Internal structure of testis*

Gamete production in the ovaries

Female reproductive system

In females, ova develop by meiosis from germline cells in the ovaries, as shown in **Figure 2.3**. The ovaries contain immature ova in various stages of development. Each developing ovum is surrounded by a follicle, as shown in **Figure 2.4**, that protects it and secretes hormones.

The follicle is a small cellular sac which secretes oestrogen. It eventually develops into a corpus luteum, which secretes progesterone.

If an ovum is fertilised in the oviduct to produce a zygote, the zygote will then divide mitotically. It moves down the oviduct aided by peristaltic waves and the action of the ciliated lining towards the uterus, where it may implant into the endometrium, eventually developing into an embryo.

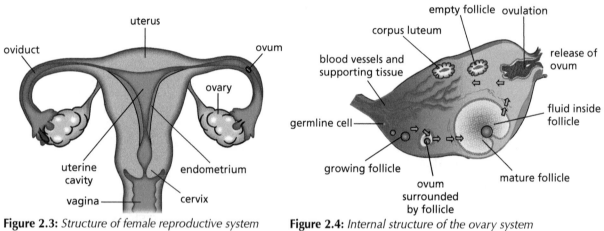

Figure 2.3: *Structure of female reproductive system* **Figure 2.4:** *Internal structure of the ovary system*

Fertilisation

Mature ova are released into the oviduct where they may be fertilised by sperm to form a zygote **(Figure 2.5)**.

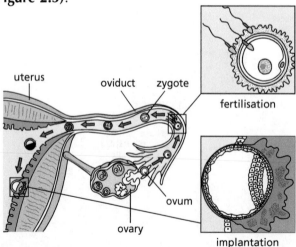

Figure 2.5: *Fertilisation in the oviduct*

TOP TIP

Contractions of the female reproductive tract by the action of hormones help move sperm towards the oviduct.

Hormonal control of reproduction

Hormonal influence on puberty

Puberty results in several changes which increase the ability of an individual to reproduce. Some of these secondary sexual changes in males have been mentioned earlier. In females they involve the development of the reproductive organs and the start of the **menstrual cycle**. At this time, the **hypothalamus** in the brain, shown in **Figure 2.6**, secretes a releaser hormone which triggers the **pituitary gland**. The pituitary gland then secretes **follicle-stimulating hormone (FSH)**, **luteinising hormone** or **interstitial cell-stimulating hormone (ICSH)**.

These hormones trigger the onset of puberty and bring about the menstrual cycle in women and sperm production in men.

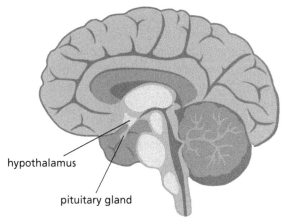

Figure 2.6: *Brain showing positions of hypothalamus and pituitary gland*

Hormonal control of sperm production

In males the hypothalamus secretes a releaser hormone which stimulates the pituitary gland to release both follicle-stimulating hormone and interstitial cell-stimulating hormone (**Figure 2.7**). These two hormones act on the testis by triggering the release of testosterone and the production of sperm cells. The secretion of the hormones is under **negative feedback control** where a high blood level of testosterone inhibits both follicle-stimulating hormone and interstitial cell-stimulating hormone production by the pituitary gland. This leads to a decrease in the blood level of testosterone, which then stimulates the pituitary gland to start secreting more follicle-stimulating hormone and interstitial cell-stimulating hormone.

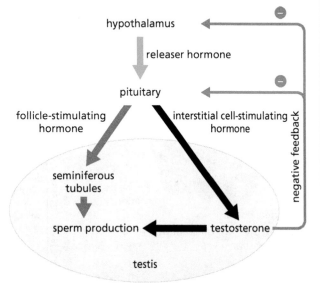

Figure 2.7: *Negative feedback involved in the production of sperm*

TOP TIP

Remember: Testosterone also activates the prostate gland and seminal vesicles.

TOP TIP

There are several bodily functions which are regulated by negative feedback, such as temperature control and blood sugar levels.

Hormonal control of the menstrual cycle

The menstrual cycle

The menstrual cycle takes approximately 28 days, with the first day of **menstruation** regarded as day 1 of the cycle. As shown in **Figure 2.8**, increasing levels of follicle-stimulating hormone secreted by the pituitary gland promotes the development of ovarian follicles and the secretion of oestrogen. Increasing levels of luteinising hormone, also secreted by the pituitary gland, causes ovulation and the development of the corpus luteum, which in turn secretes progesterone.

There are two phases of the menstrual cycle, called the **follicular phase** and the **luteal phase**. These phases, shown in **Figure 2.9**, have profound effects on the ovaries.

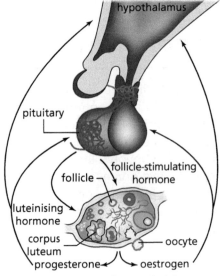

Figure 2.8: *Hormonal control of the menstrual cycle*

The follicular phase

In the follicular phase (**Figure 2.10**), follicle-stimulating hormone secreted by the pituitary gland causes the maturation of a follicle and the secretion of oestrogen from the ovaries. The increased level of oestrogen restores the wall of the endometrium after the bleeding associated with the last menstrual cycle. Thus it prepares the endometrium for implantation. It also decreases the density of the mucus at the cervix, making it easier for sperm to swim through. As the oestrogen levels rise then peak, the pituitary gland produces a surge of luteinising hormone, starting ovulation around the fourteenth day of the menstrual cycle.

TOP TIP

Releaser hormone from the hypothalamus stimulates the release of follicle-stimulating, interstitial cell stimulating and luteinising hormones.

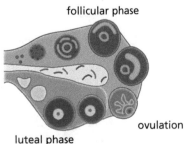

Figure 2.9: *Two different phases of the menstrual cycle*

TOP TIP

Remember: **Ovulation** is the release of an ovum from a follicle in the ovary. It usually occurs around the mid-point of the menstrual cycle.

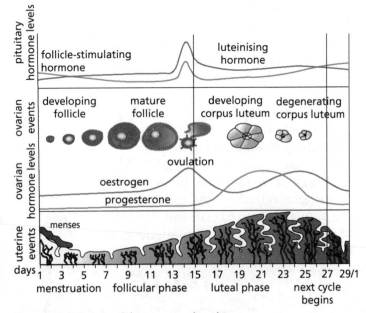

Figure 2.10: *Events of the menstrual cycle*

The luteal phase

In the luteal phase, luteinising hormone promotes the development of a follicle into a corpus luteum. The corpus luteum secretes progesterone, a hormone which promotes the further development and vascularisation of the endometrium. This prepares the endometrium for implantation if fertilisation occurs. The density of the mucus at the cervix also increases. During this stage, the increased levels of progesterone and oestrogen have a negative feedback on the pituitary gland, which decreases the secretion of both luteinising hormone and follicle-stimulating hormone so that no new follicles develop.

If the ovum is not fertilised, the corpus luteum usually degenerates so that the levels of progesterone fall. This in turn triggers the next menstrual cycle to start, with the loss of tissue from the endometrium and 30–40 cm^3 of blood (**Figure 2.11**).

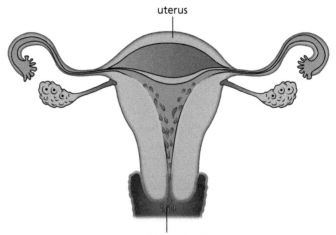

uterus

uterine lining and unfertilised egg are shed during menstruation

Figure 2.11: *Menstrual flow of blood*

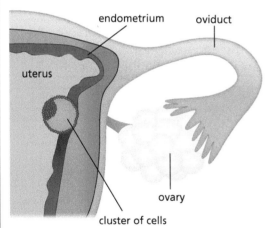

endometrium   oviduct

uterus

ovary

cluster of cells

Figure 2.12: *Development of zygote into blastocyst*

If fertilisation does take place, the onset of the next menstrual cycle is prevented by a hormone secreted by the zygote which mimics the effect of luteinising hormone. When this happens the corpus luteum is not able to degenerate but is maintained and continues to produce progesterone. The zygote develops into a cluster of cells (**Figure 2.12**), which becomes implanted around seven days after fertilisation in the endometrium to develop into an embryo. The continued production of progesterone is taken over by the placenta.

Hormones associated with the menstrual cycle

Hormone	Where secreted	Effect of hormone
releaser hormone	hypothalamus	stimulates release of follicle-stimulating hormone (FSH), luteinising hormone (LH) or interstitial cell-stimulating hormone (ICSH) this triggers the onset of puberty
follicle-stimulating hormone	pituitary gland	FSH stimulates the development of a follicle and the production of oestrogen by the follicle in the follicular phase
luteinising hormone	pituitary gland	surge in luteinising hormone triggers ovulation while a lack of luteinising hormone leads to degeneration of the corpus luteum the subsequent drop in progesterone leads to menstruation
oestrogen	ovary and follicle	stimulates proliferation of the endometrium in preparation for implantation oestrogen affects the consistency of cervical mucus making it more easily penetrated by sperm peak levels of oestrogen stimulate a surge in secretion of LH. This surge in LH triggers ovulation
progesterone	ovary and corpus luteum	progesterone promotes further development and thickening of the endometrium, preparing it for implantation if fertilisation occurs

The biology of controlling fertility

Infertility treatments

Fertility may be affected by a number of factors, shown in **Figure 2.13**, acting singly or in combination, resulting in an inability to reproduce successfully.

Treatment of infertility is based on an understanding of fertility and the reproductive process.

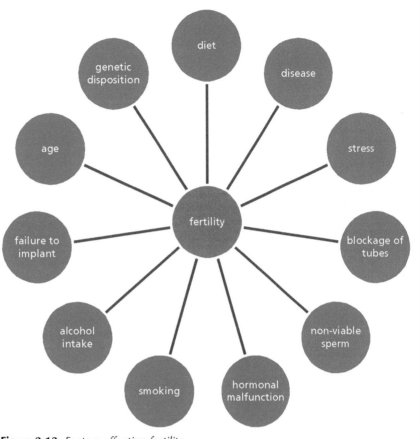

Figure 2.13: *Factors affecting fertility*

Fertile periods

TOP TIP

Remember: Women are only fertile for a few days in each menstrual cycle.

In males, the blood levels of follicle-stimulating hormone and interstitial cell-stimulating hormone remain within narrow limits. Consequently, the concentration of testosterone does not change much and therefore sperm production remains consistent, resulting in an ability to reproduce at any time. This is called **continuous fertility**.

In contrast, a female's reproductive potential is linked to a cycle of events meaning that she can only conceive within three to five days of ovulation, known as the **fertile period** (**Figure 2.14**). During ovulation the body temperature drops slightly then increases. Females have a fertility which is **cyclical**.

Sperm are able to survive for several days after intercourse, which means that they can potentially still fertilise an ovum if ovulation takes place while the sperm are still present and viable.

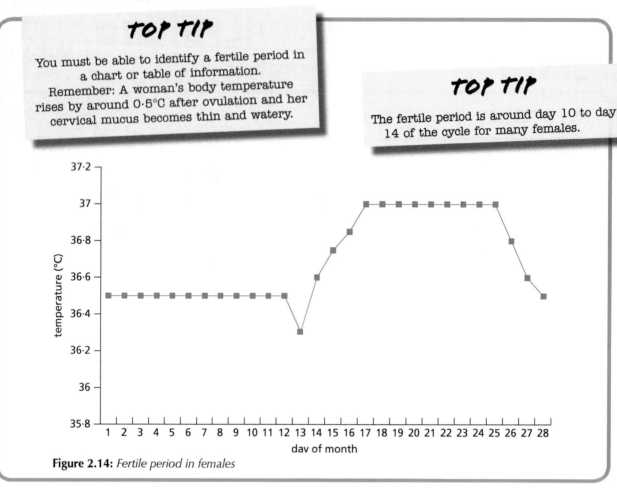

Figure 2.14: *Fertile period in females*

Treatments for infertility

1. Stimulating ovulation: ovulation is stimulated by drugs that prevent the negative feedback effect of oestrogen on FSH secretion. Different ovulatory stimulating drugs work by mimicking the action of the hormones FSH and LH, causing **super ovulation**. Multiple births can occur as more than one ovum is released and may potentially be fertilised. Sometimes the ova produced as a result of super ovulation can be collected for in vitro fertilisation (IVF) programmes.

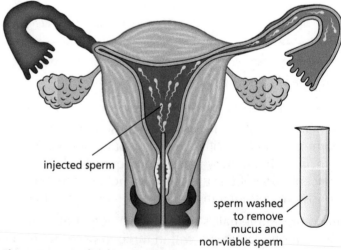

injected sperm

sperm washed to remove mucus and non-viable sperm

Figure 2.15: *Artificial insemination*

2. **Artificial insemination**, shown in **Figure 2.15**, introduces sperm into the female reproductive tract using an artificial means instead of sexual intercourse. Samples of semen are collected over time. Artificial insemination is very useful when a male

has a low sperm count as many samples can be combined to increase sperm number. If a male partner is **sterile** the sperm can be provided by using semen obtained from a voluntary donor bank.

3. A relatively new technique, shown in **Figure 2.16**, called **intra-cytoplasmic sperm injection (ICSI)** uses only one sperm to fertilise one ovum. It is used where there is an issue with the number of sperm and/ or if sperm are defective. A glass needle is used to inject the sperm head, containing the nucleus, directly into the ovum to achieve fertilisation.

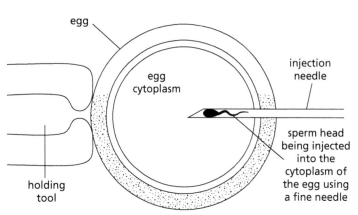

Figure 2.16: *intra-cytoplasmic sperm injection (ICSI)*

4. **In vitro fertilisation (IVF)**, shown in **Figure 2.17**, is a procedure in which ova are removed from ovaries by surgery following hormone stimulation. They are mixed with sperm in a culture dish in a laboratory. The fertilised eggs are then incubated in nutrient medium until they have formed at least eight cells and are then inserted into the uterus for implantation. Some of the fertilised eggs that are not used for IVF may, with consent, be used for stem cell research. A **pre-implantation genetic diagnosis (PGD)** can be carried out to check for any single gene disorders or chromosomal abnormalities. Unused embryos can be frozen and stored then used later.

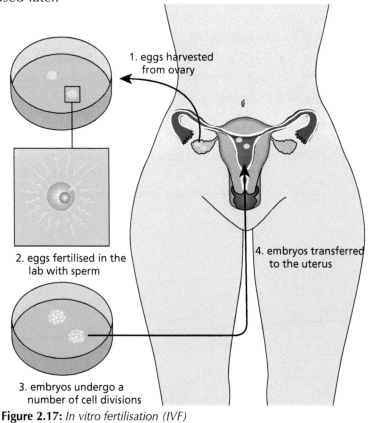

Figure 2.17: *In vitro fertilisation (IVF)*

Physical and chemical methods of contraception

Contraception is the deliberate prevention of conception and can be either physical or chemical. Physical methods prevent sperm and egg meeting. Chemical methods use synthetic hormones which mimic the effects of those produced naturally.

Physical methods of contraception	
Method	*Description*
intra-uterine device (IUD)	small copper or plastic device inserted into the uterus to prevent implantation
barriers	physical methods of preventing sperm and egg meeting such as a **condom, cervical cap** and **diaphragm.**
sterilisation	surgical procedure, usually irreversible, in which the sperm ducts are cut and sealed preventing sperm from entering semen

Chemical methods of contraception	
Method	*Description*
oral contraceptive pill	synthetic versions of the hormones progesterone and oestrogen which have the same negative feedback properties on the secretions of the pituitary gland and prevent the release of FSH and LH from the pituitary gland
morning after pill	used in circumstances when no other method of contraception was available at the time of sexual intercourse or if the original method of contraception failed
	contains very high levels of the synthetic forms of progesterone and oestrogen preventing ovulation and implantation
	also causes thickening of the mucus around the cervix making it difficult for sperm to penetrate
progesterone-only (mini) pill	contains only the synthetic version of progesterone causing negative feedback on the pituitary gland as well as an increased thickening of the cervical mucus

Antenatal and postnatal screening

Antenatal screening

The health of a mother-to-be and her developing fetus and baby can be monitored by a number of techniques. Antenatal screening is a method of detecting potential issues with an embryo or fetus before birth. For example, where there is a family history of a genetic condition. The number of conditions which can be detected before birth is increasing all the time. The majority of the techniques used are non-invasive and are used during the first six months of pregnancy. Antenatal testing can reveal the sex of the developing embryo as well as its blood group, identify the risk of a genetic disorder and allow prenatal diagnosis of disease. The screening will also indicate the delivery date.

Ultrasound imaging

Ultrasound imaging, shown in **Figure 2.18**, makes use of high-frequency sound waves passed through the abdominal wall to create an image of the fetus in the uterus. Ultrasound scanning is a technique that was developed in Scotland. Currently, women are given two ultrasound scans during pregnancy. A dating scan takes place between 8 and 14 weeks. It determines the stage of the pregnancy and can be used to calculate when the baby will be born. Dating scans are

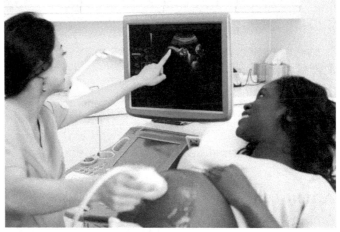

Figure 2.18: *Ultrasound imaging*

used alongside biochemical tests for chemical markers such as proteins, which vary in concentration during pregnancy. An anomaly scan takes place between 18 and 20 weeks, and may detect serious physical abnormalities in the fetus.

Biochemical testing of blood and urine

As pregnancy proceeds, the levels of proteins produced by the placenta can be tested by taking a blood sample, as shown in **Figure 2.19**.

Blood and urine tests are carried out routinely throughout pregnancy to monitor the concentrations of marker chemicals. For example, unusually low or atypical levels of a particular protein marker molecule can help identify a fetus with Down's syndrome.

As levels of chemicals vary during pregnancy, great care needs to be taken with regard to the timing of these tests. Measuring a chemical at the wrong time could lead to a false positive result. Dating scans may help to ensure appropriate timing of biochemical tests and decrease the chance of false results.

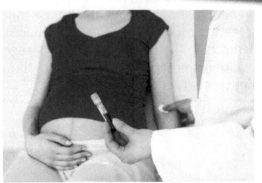

TOP TIP

An unusual concentration of a marker chemical leads to diagnostic testing being carried out to investigate whether the fetus has a medical condition.

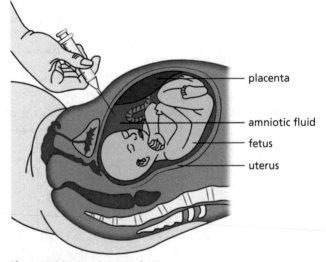

Figure 2.19: *Blood sample to be used for biochemical testing*

Diagnostic testing

Diagnostic tests involve taking a sample of either the placenta, the amniotic fluid or fetal blood. The samples are then sent away and examined in a laboratory for chromosomal or genetic abnormalities. The two main diagnostic tests are **amniocentesis** and **chorionic villus sample (CVS)**.

Amniocentesis

Amniocentesis, shown in **Figure 2.20**, is used to diagnose potential birth abnormalities by examining cells found in the amniotic fluid. This takes place between the fifteenth and twentieth weeks of pregnancy and is offered to women over the age of 35 or who have already had a blood test indicating potential abnormalities. A small sample of the **amniotic fluid** which surrounds the fetus is removed using a long thin needle inserted through the abdominal wall. The amniotic fluid contains cells shed by the fetus which are used for genetic analysis. This technique can help detect possible defects in the development of the brain or spinal cord.

placenta

amniotic fluid

fetus

uterus

Figure 2.20: *Amniocentesis*

Chorionic villus sampling

Chorionic villus sampling, shown in **Figure 2.21**, is used to diagnose potential congenital abnormalities by examining cells from the placenta. This takes place between the tenth and thirteenth weeks of pregnancy and is offered to women who have a family history of a genetic condition or have already had a blood test indicating potential abnormalities. Cells containing the same genetic material as the developing fetus are removed from the placenta. A long flexible tube is inserted through the vagina into the cervix. It is then guided towards the placenta and the chorionic villi, shown in **Figure 2.22**.

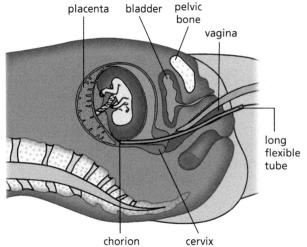

Figure 2.21: *Chorionic villus sampling*

A small sample of tissue is taken from the placenta and used for genetic analysis. Unlike amniocentesis, this technique cannot help detect possible defects in the development of the brain or spinal cord.

Both types of diagnostic testing carry a risk of inducing a miscarriage. The risk in amniocentesis is approximately 1% after the fifteenth week of pregnancy. The risk of miscarriage in chorionic villus sampling is estimated to be 1–2%. Both tests have advantages and disadvantages. Chorionic villus sampling can be carried out earlier in a pregnancy than amniocentesis but the former carries a higher risk of miscarriage occurring.

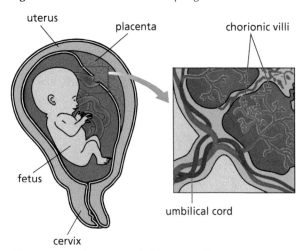

Figure 2.22: *Placenta and chorionic villi*

Cells taken by amniocentesis and/or chorionic villus sampling can be used to produce a representation of the chromosomes present in the cells of a baby, known as a karyotype. A person's karyotype is a visual record of their chromosomes arranged as homologous pairs.

Any abnormalities in terms of the numbers and/or shapes of the chromosomes can help diagnose genetic conditions like Down's syndrome (**Figure 2.23**). In deciding to go ahead with karyotyping or any other tests, the element of risk to the developing baby will be assessed. Individuals concerned will also be advised to think carefully about the options available to them if any test result is returned positive.

Figure 2.23: *Karyotype showing extra chromosome number 21*

Inheritance patterns

Patterns of inherited conditions can be shown by collecting data over several generations of a family to produce pedigree charts which reveal both the phenotypes and genotypes of family members. This information is useful if a family is known to be at high risk of a genetic condition such as albinism, Huntington's disease, thalassaemia, haemophilia or muscular dystrophy.

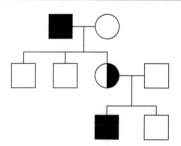

A pedigree chart, as shown in **Figure 2.24**, shows how a genetic trait is passed through several generations of a family. This is useful as a predictor of how likely an individual is to have or to develop such a genetic trait. It can be used to track how a genetic condition (particularly if it is an unusual one) is expressed over several generations, and provide increased knowledge on which to base decisions about having children.

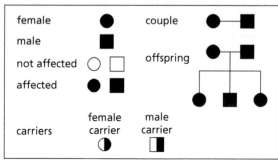

Figure 2.24: *Typical pedigree chart showing how a trait is inherited*

TOP TIP

You need to be able to draw, analyse and interpret genetic family histories (pedigree charts) over three generations in order to follow patterns of inheritance of disease.

Four important inheritance patterns can be shown using pedigree charts.

Defective recessive autosomal inheritance

A disorder such as albinism (**Figure 2.25**) which is caused by a defective recessive autosomal allele:

- is expressed relatively rarely
- usually skips generations
- affects both males and females in equal numbers
- requires the affected individual to be homozygous recessive
- can be carried by a heterozygous individual
- can result from two unaffected parents who are heterozygous.

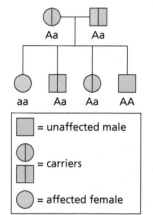

Figure 2.25: *Pedigree of defective recessive autosomal inheritance*

Defective autosomal dominant allele

A disorder such as Huntington's disease (**Figure 2.26**) which is caused by a defective autosomal dominant allele:

- affects both males and females in equal numbers
- means anybody affected will have an affected parent
- no longer appears in future generations if a branch of the pedigree does not show the disorder
- results in all non-affected individuals being homozygous recessive
- means anybody affected is either double dominant or heterozygous.

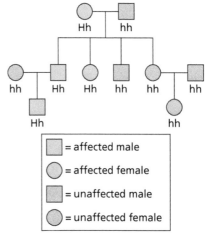

Figure 2.26: *Pedigree of defective dominant autosomal inheritance*

Autosomal defective allele

A disorder such as sickle cell disease (**Figure 2.27**) which is caused by an autosomal defective allele:

- affects both males and females in equal numbers
- in the homozygous state produces the maximum expression of the phenotype
- is rarely expressed maximally as the two alleles present are incompletely dominant over each other
- in the heterozygous state produces a reduced expression of the phenotype.

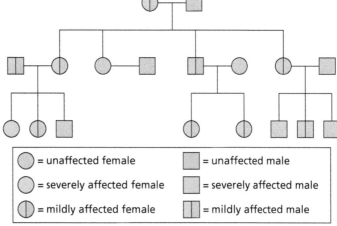

Figure 2.27: *Pedigree showing inheritance of sickle-cell anaemia*

Defective sex-linked recessive allele

A disorder such as haemophilia (**Figure 2.28**) which is caused by a defective sex-linked recessive allele:

- affects many more males than females
- is not transmitted to a male from his affected father
- needs an individual to be homozygous if they are female
- will be expressed in a male who has one copy of the defective allele
- will not be expressed in a homozygous or heterozygous female or a male who has the normal allele
- means usually the mother of an affected male is, herself, unaffected
- means all daughters of affected fathers will either be carriers or be affected themselves.

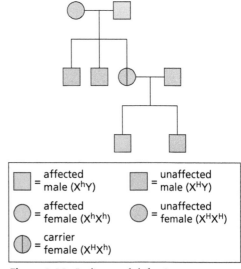

Figure 2.28: *Pedigree of defective sex-linked inheritance*

Postnatal screening

Postnatal screening is the use of diagnostic testing to check for any abnormalities in a baby after it is born.

One important use of screening as a diagnostic test is to check for an error in metabolism called **phenylketonuria** (PKU), which is caused by an autosomal recessive allele. Babies born in the UK are tested for this condition by taking a blood sample from the heel as shown in **Figure 2.29**. In PKU a substitution mutation means that the enzyme which converts phenylalanine to tyrosine does not function.

If a baby is diagnosed with phenylketonuria, then a restricted diet is prescribed. This diet will contain low levels of the amino acid phenylalanine. The diet will avoid certain foods such as cereals and potatoes. Providing the diet is followed then the brain function of the baby will develop normally. The restricted diet is usually continued into adulthood.

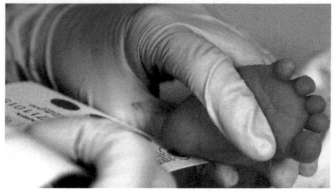

Figure 2.29: *Postnatal screening for phenylketonuria*

The structure and function of arteries, capillaries and veins

Circulatory system

Transport of materials around the body depends on the circulatory system. Blood is pumped from the heart through the arteries to the capillaries then to the veins and eventually returns back to the heart, as shown in **Figure 2.30**. Blood pressure is a term used to describe the force of the blood exerted on vessel walls. Blood pressure decreases as blood moves around the body, reaching its lowest when it returns to the right atrium. This drop in pressure is due to friction occurring between the circulating blood and the walls of the vessel it is being transported in.

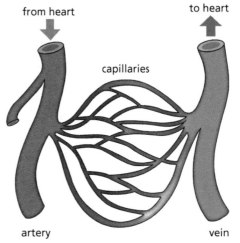

Figure 2.30: *Artery and vein connected by capillaries*

Within the blood vessels is a space called the **central lumen** through which blood travels. The central lumen is lined with special cells which form a smooth layer called the **endothelium**. The endothelium of each of the three main types of blood vessel is surrounded by layers of tissue. The composition of the tissue in the vessel walls of the blood vessels differs and this difference is related to their function.

Arteries

Arteries are vessels which carry blood away from the heart. Their walls, shown in **Figure 2.31**, are adapted to withstand the very high pressure of blood which is a result of the muscular contractions of the heart.

The walls of arteries consist of three layers. Two of these, **connective tissue** and smooth muscle layers, contain **elastic fibres**. The elastic walls of the arteries stretch and recoil (pulsate) to accommodate the surge of high pressure blood following contraction of the heart.

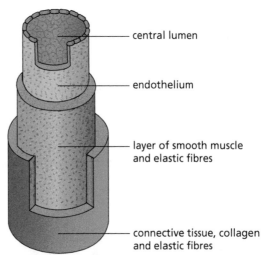

Figure 2.31: *Walls of an artery*

The smooth muscle in the arterial wall is capable of relaxing (**vasodilation**) or contracting (**vasoconstriction**) thus controlling blood flow in response to different demands, as shown in **Figure 2.32**. For example, in cold weather, the blood vessels near the skin can vasoconstrict causing less radiative heat loss, while strenuous exercise can cause these blood vessels to vasodilate to allow more radiative heat loss.

> **TOP TIP**
>
> Radiative heat loss is linked to the autonomic nervous system studied later in this area.

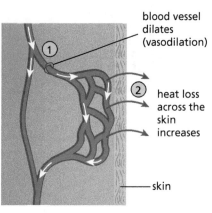
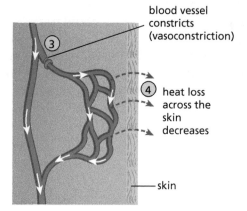

1. blood vessel dilation results in increased blood flow toward the surface of the skin

2. increased blood flow beneath the epidermis results in increased heat loss (blue arrows)

3. blood vessel constriction results in decreased blood flow toward the surface of the skin

4. decreased blood flow beneath the epidermis results in decreased heat loss

Figure 2.32: *Vasodilation and vasoconstriction in the skin*

Veins

Veins are blood vessels which return blood to the heart. Veins have an outer layer of connective tissue containing elastic fibres. The layer of muscle in the wall of a vein is much thinner than that of an artery since the blood flow is under much less pressure. The central lumen of a vein is usually much larger than that of an artery to allow blood to flow more easily back to the heart. To help the one-way flow of blood back to the heart, veins also have valves, as shown in **Figure 2.33**.

> **TOP TIP**
>
> Remember: Valves prevent the backflow of blood.

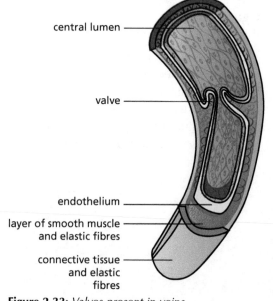

Figure 2.33: *Valves present in veins*

Capillaries

Blood travels through arteries towards veins via a dense network of tiny capillaries, as shown in **Figure 2.34**.

The tiny arteries and veins are called **arterioles** and **venules**, respectively.

Exchange of materials such as nutrients, gases, fluids and wastes takes place across the thin walls of the capillaries.

> **TOP TIP**
>
> The walls of a capillary are only one cell thick.

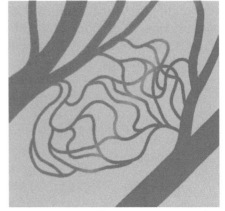

Figure 2.34: *Capillaries connect arterioles to venules*

Tissue Fluid

When blood arrives at a capillary bed, the pressure forces some of the plasma (**Figure 2.35**) out through the thin walls into the space outside. This is known as **pressure filtration**.

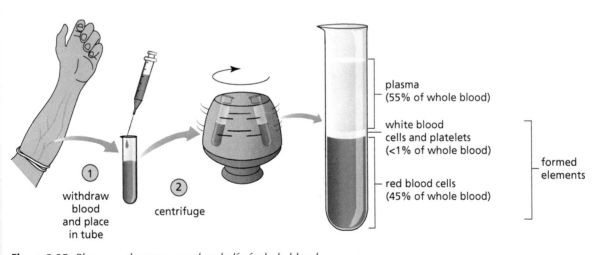

Figure 2.35: *Plasma makes up more than half of whole blood*

> **TOP TIP**
>
> Plasma contains many dissolved substances such as glucose, amino acids, carbon dioxide, oxygen, urea, vitamins, minerals, hormones and antibodies.

The liquid forced out of the capillaries is called **tissue fluid** and is similar to plasma but contains no protein molecules (**Figure 2.36**). Tissue fluid allows exchange of materials between capillaries and tissue cells and supplies them with glucose, oxygen and other substances. Wastes produced by cellular metabolism such as carbon dioxide diffuse out of the cells into the tissue fluid and are excreted from the body. Much of the tissue fluid returns to the blood by osmosis.

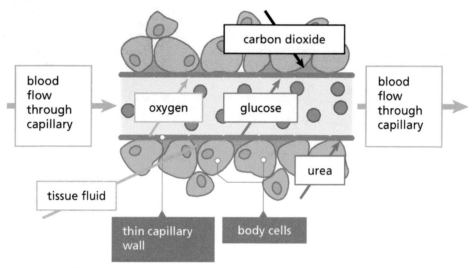

Figure 2.36: *Tissue fluid formation*

Not all the tissue fluid returns back to the capillary. Instead the excess tissue fluid enters the lymphatic system (**Figure 2.37**) via lymphatic vessels. These are unusual in that they are closed at one end. **Lymph** collected from all the lymphatic vessels eventually re-enters the circulatory system.

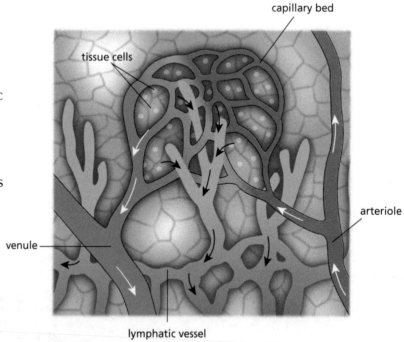

Figure 2.37: *The lymphatic system*

TOP TIP

Tissue fluid and blood plasma are similar in composition. However, unlike plasma, tissue fluid does not contain proteins, which are too large to be pressure-filtered through the capillary walls.

The structure and function of the heart

Cardiac output and its calculation

Blood is pumped through the circulatory system by the heart. Blood enters the heart by the two atria and leaves via the two ventricles, as shown in **Figure 2.38**. The right side pumps deoxygenated blood while the left side pumps oxygenated blood. The same volume of blood is pumped by the right ventricle through the pulmonary artery as is pumped by the left ventricle through the aorta. Valves prevent backflow so that the blood travels in one direction.

> ### TOP TIP
> The average body contains about 5 litres of blood, which means that all of our blood is pumped through our hearts about once every minute.

The **cardiac output (CO)** is the name given to describe the volume of blood pumped through each ventricle per minute. Cardiac output is determined by heart rate and **stroke volume (SV)**.

cardiac output (CO) = heart rate (HR) × stroke volume (SV)

The cardiac output of an untrained adult with a resting heart rate of 72 beats per minute and a stroke volume of 70 cm³/minute will be 5040 cm³/minute. A highly trained athlete's cardiac output can be eight times this figure, due to the efficiency of their heart function.

> ### TOP TIP
> When you exercise, cardiac output increases to meet the increasing demand for oxygen, glucose and the removal of carbon dioxide.

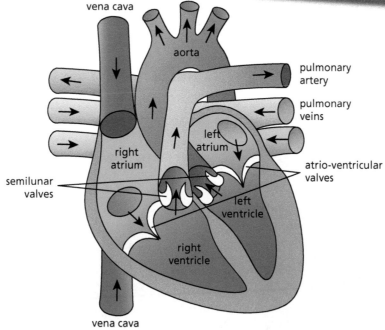

Figure 2.38: *Structure of heart, associated blood vessels and valves.*

The cardiac cycle

Each single beat of the heart, lasting on average about 0·8 seconds, is made up of a sequence of highly co-ordinated events known as the cardiac cycle, consisting of two phases (**Figure 2.39**):

1. Relaxation phase known as diastole when the heart muscle is relaxed and the heart fills up with blood.

2. Contraction phase known as systole when the heart muscle is contracted and blood is forced from the atria (atrial systole) into the ventricles and from the ventricles (ventricular systole) out of the heart.

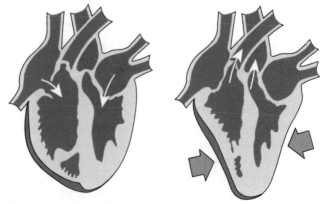

Figure 2.39: *Two phases of the cardiac cycle*

Linked to these two different phases is the action of the valves in the heart. The atrio-ventricular valves (AV) are found between the two atria and the two ventricles. The semilunar valves (SL) are located at the exit of the ventricles (**Figure 2.40**).

During diastole, blood returning to the atria flows into the ventricles.

Contraction of the atria during atrial systole pumps the remainder of the blood through the AV valves and into

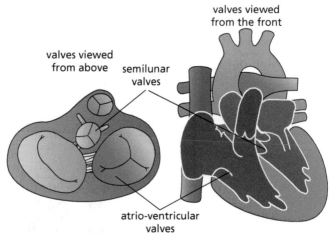

Figure 2.40: *Location of AV and SL valves in the heart*

the ventricles. The contraction of the ventricles, ventricular systole, closes the AV valves and pumps the blood out of the heart through the SL valves to the pulmonary artery and aorta, as shown in **Figure 2.41**.

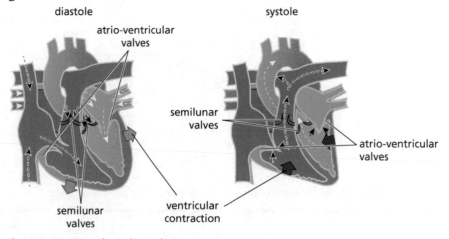

Figure 2.41: *Diastole and systole*

In diastole, the higher blood pressure in the arteries leaving the heart closes the semilunar valves. This closing and opening of valves can be heard with a stethoscope. The entire sequence of events which make up the cardiac cycle, as shown in **Figure 2.42**, takes place in less than one second.

This consists of contraction and relaxation of the heart muscle, systole and diastole, and the smooth operation of the valves to ensure the one-way flow of blood in and out of the heart.

As the valves open and close in pairs they make a characteristic sound known as 'lubb' and 'dub' (**Figure 2.43**).

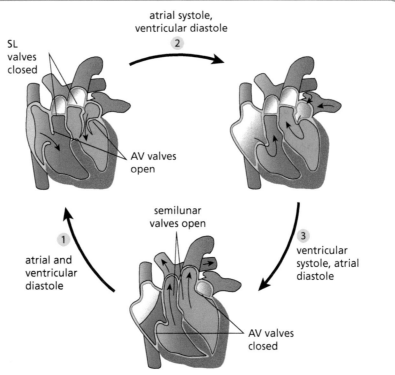

Figure 2.42: *Cardiac cycle*

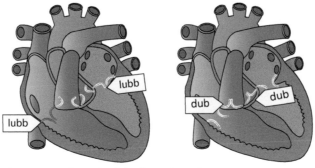

Figure 2.43: *Sounds of the heart in action*

TOP TIP

The 'lubb' and 'dub' sounds are what are picked up by a stethoscope against the chest wall.

The structure and function of the cardiac conducting system

The heart muscle is unusual in that it can contract without an external nervous supply. This ability of heart muscle cells to generate their own stimulation ensures the whole heart contracts in a rhythmic and highly co-ordinated way using its own conducting system, as shown in **Figure 2.44**.

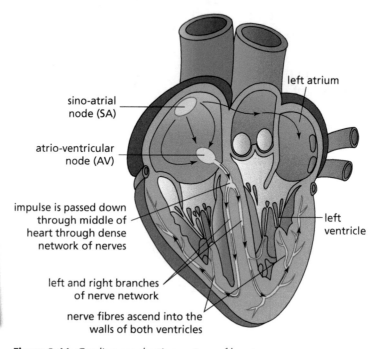

TOP TIP

Although the heart can set its own pace, the actual rate of heartbeat is highly variable and external nerve impulses can speed up or slow down this rate according to demands.

Figure 2.44: *Cardiac conducting system of heart*

The origin of the heartbeat is in the heart itself. The cells of the **sino-atrial node** (SAN) or pacemaker, located in the wall of the right atrium, set the rate at which the heart contracts. These cells are auto-rhythmic. This means that unlike other body cells they have an intrinsic beating property and show spontaneous excitation.

The timing of cardiac muscle cell contraction is controlled by electrical impulses from the SAN spreading through the muscle cells of both atria and causing atrial systole. The impulses are then picked up by the **atrioventricular node** (AVN), located in a central position at the base of the atria. Impulses from the AVN travel down conducting fibres in the central wall of the heart and then up through the walls of the ventricles, simultaneously causing contraction in both ventricles (ventricular systole).

The arrangement of the nerves from the atrioventricular node causes the ventricles to contract from the base upwards, squeezing out the blood into the aorta or pulmonary artery.

Electrocardiograms

Impulses in the heart generate currents that can be detected by electrodes placed on the skin. The pattern of these when displayed as a trace is known as an **electrocardiogram** (ECG). This is shown in **Figure 2.45**.

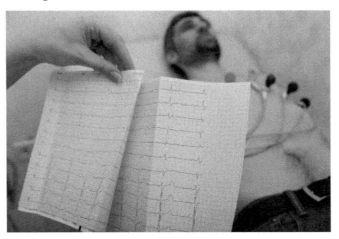

Figure 2.45: *Electrical activity of heart generating an electrocardiogram*

A normal ECG consists of three connected waves identified by the letters P, QRS and T.

These waves are shown in **Figure 2.46** and described here:

- P wave – right and left atria are stimulated to contract. The P wave lasts less than 0·12 seconds
- QRS wave – right and left ventricles are stimulated to contract. The QRS wave lasts less than 0·10 seconds
- T wave – recovery period of both right and left ventricles. The T wave lasts about 0·25 seconds. Ventricular diastole is represented by the time between the end of the T wave and the start of the QRS complex.

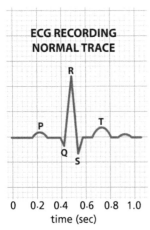

Figure 2.46: *An ECG trace showing the different waves produced during one cardiac cycle*

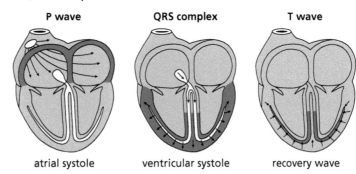

Figure 2.47: *P, QRS and T waves*

TOP TIP

Be sure you can calculate the heartbeat from an ECG trace. A normal cardiac cycle lasting 0·8 seconds means the heart is beating at 75 bpm.

Changes to the normal trace can indicate the heart is not functioning properly, as shown in **Figure 2.48**.

ventricles contracting more rapidly than normal

fast heartbeat

ventricles contracting slower than normal

slow heartbeat

ventricles contracting irregularly

irregular heartbeat

atria contracting more rapidly than normal

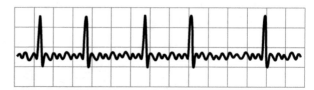

The medulla in the brain has a number of functions, one of which is to regulate the rate of the sino-atrial node (**Figure 2.49**).

Figure 2.48: *Electrocardiograms can reveal abnormal heart function*

It does this through the **antagonistic** action of the autonomic nervous system. Sympathetic nerve action increases the activity of the sino-atrial node, causing the heart to beat faster and increase the cardiac output. Sympathetic nerve fibres are over almost all of the heart muscle. Parasympathetic activity has the opposite effect, slowing down the heartbeat and decreasing the cardiac output. These actions are antagonistic to each other and are mediated by two different neurotransmitters, as shown in the table below.

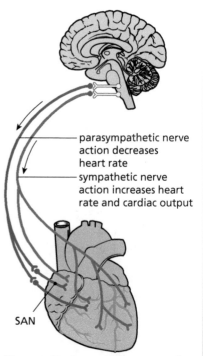

parasympathetic nerve action decreases heart rate

sympathetic nerve action increases heart rate and cardiac output

SAN

Figure 2.49: *Antagonistic action of the autonomic control of heart rate*

Division of autonomic nervous system	Neurotransmitter secreted	Effect
parasympathetic	acetylcholine	heart rate, pulse rate and cardiac output decrease
sympathetic	noradrenaline	heart rate, pulse rate and cardiac output increase

With each contraction of the left and right ventricles blood is forced into the arterial system under pressure (**Figure 2.50**).

Each contraction and relaxation gives rise to systolic and diastolic blood pressures, respectively. The maximum systolic pressure occurs when the ventricles exert maximum force on the blood, which falls to its minimum when the ventricles relax.

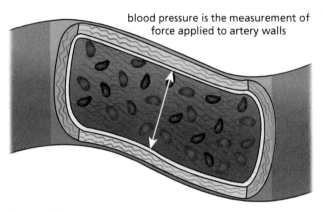

blood pressure is the measurement of force applied to artery walls

Figure 2.50: *Blood pressure in an artery*

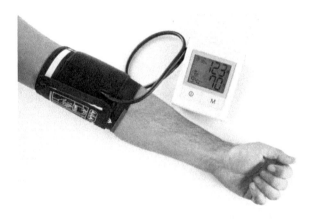

Blood pressure is measured using a digital device called a **sphygmomanometer** (**Figure 2.51**).

An inflated cuff, positioned around the upper arm, temporarily stops arterial blood flow. When the cuff is allowed to deflate gradually the blood starts to flow. This is detected by a pulse at systolic pressure (typically 120 mmHg). The blood then flows freely through the artery (and a pulse is not detected) at diastolic pressure (typically 80 mmHg).

Figure 2.51: *A sphygmomanometer measures blood pressure*

A young person has a typical blood pressure reading of 120/80 mmHg.

Prolonged high blood pressure is known as **hypertension** and is a major risk factor for coronary heart disease and strokes (**Figure 2.52**).

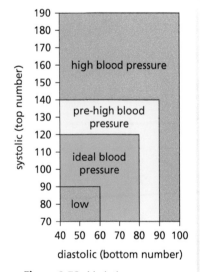

Figure 2.52: *Variations in blood pressure*

Pathology of cardiovascular disease (CVD)

Atherosclerosis

Depending on a number of factors such as age, lifestyle, diet, smoking and genetics, arteries may have an accumulation of a fatty material which includes cholesterol, calcium and fibrous material forming an **atheroma**, as shown in **Figure 2.53**.

The thickened artery loses its natural elasticity as the plaque beneath the endothelium grows. Blood flow becomes restricted due to the reduced diameter of the lumen. These factors combine and cause an increase in blood pressure.

Atherosclerosis is thus the root cause of many cardiovascular diseases (CVD), such as angina, peripheral vascular disease, heart attack and stroke.

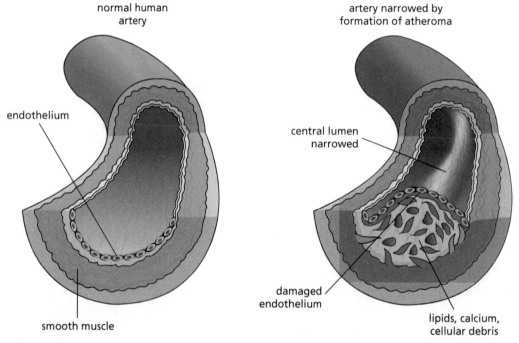

Figure 2.53: *Formation of atheroma in an artery*

Thrombosis

As an atheroma builds up it becomes a site for the formation of a blood clot called a **thrombus**. Blood clotting is the process which prevents bleeding when a blood vessel is damaged. A series of reactions occurs to stop bleeding by forming a clot to act as a plug. The process starts with **prothrombin**, an inactive enzyme, being converted into its active form, **thrombin**. Thrombin converts **fibrinogen** into **fibrin**, forming meshwork which traps blood platelets and blood cells creating a clot (**Figure 2.54**).

There is a risk the thrombus will be dislodged and become mobile, forming an **embolus** which can potentially block blood flowing to the heart and resulting in a **myocardial infarction (MI)**.

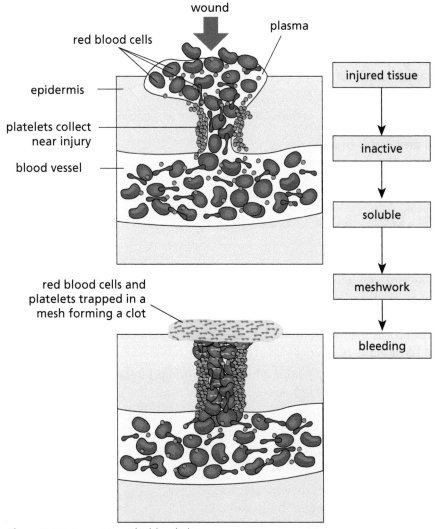

Figure 2.54: *Formation of a blood clot*

TOP TIP

If the endothelium of a blood vessel is damaged, this smooth surface is lost which can provoke clot formation.

Causes and effects of peripheral vascular disorders

Peripheral vascular disease (PVD) occurs when the damage caused by narrowing of arteries takes place distant from the heart or brain. One of the most common forms of peripheral vascular disease is **deep vein thrombosis (DVT)**, when a blood clot forms in a deep vein, usually in a leg. Pain results as the leg muscles have a decreased supply of oxygen. Sometimes, part of the clot becomes detached to form an embolus that can break off and result in a pulmonary embolism, as shown in **Figure 2.55**.

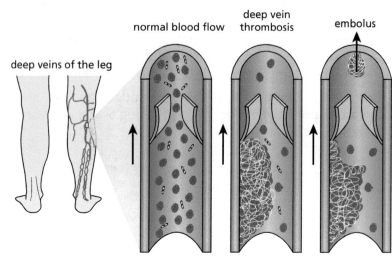

Figure 2.55: *Formation of an embolus*

Control of cholesterol levels

Cholesterol is an insoluble lipid which forms part of an animal cell membrane. It is used to make the sex hormones – testosterone, oestrogen and progesterone. Cholesterol is made in all cells but a quarter of all production takes place in the liver. It is known that when a person consumes a diet high in saturated fats cholesterol levels increase in the blood. Because cholesterol is insoluble in the blood, it is transported by **lipoproteins**, which are soluble in the blood and act as carriers. There are two types of lipoprotein carriers:

1. **Low-density lipoprotein (LDL)**, which is considered to be bad because it promotes the formation of atheroma. High levels of LDL increase the risk of cardiovascular disease.

2. **High-density lipoprotein (HDL)**, which is considered to be good because it helps remove LDL from the arteries by acting as a scavenger molecule, carrying the LDL back to the liver to be broken down and excreted. Up to one third of cholesterol is carried by HDL, which is believed to protect against heart attack and stroke.

LDL transports cholesterol to cells which have special receptors on their surface. The number of LDL receptors is directly affected by how much cholesterol is in each cell. Once a cell has sufficient cholesterol a negative feedback system inhibits the synthesis of new LDL receptors and LDL circulates in the blood where it may deposit cholesterol in the arteries forming atheromas. A higher ratio of HDL to LDL will result in lower blood cholesterol and a reduced chance of atherosclerosis.

> ### TOP TIP
> LDL transports cholesterol to body cells. HDL transports excess cholesterol from the body cells to the liver for elimination. This prevents accumulation of cholesterol in the blood.

There are great health benefits, such as elevated HDL levels, associated with regular physical activity and a low fat diet. Current advice from health practitioners aims to encourage the reduction of dietary fat intake and replace saturated with unsaturated fats.

Statins **(Figure 2.56)** are commonly prescribed drugs that reduce blood cholesterol by inhibiting the synthesis of cholesterol by liver cells.

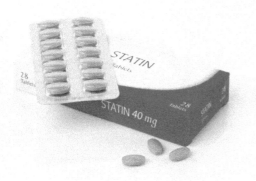

TOP TIP

There are several ways to reduce the level of LDLs and increase the level of HDLs. These include eating less animal fat and eating more fats found in fish, nuts, plant spreads and oils, eating more fruit, wholegrain food, beans and other legumes, taking regular exercise and giving up smoking.

Figure 2.56: *Statins are widely prescribed to control high blood cholesterol*

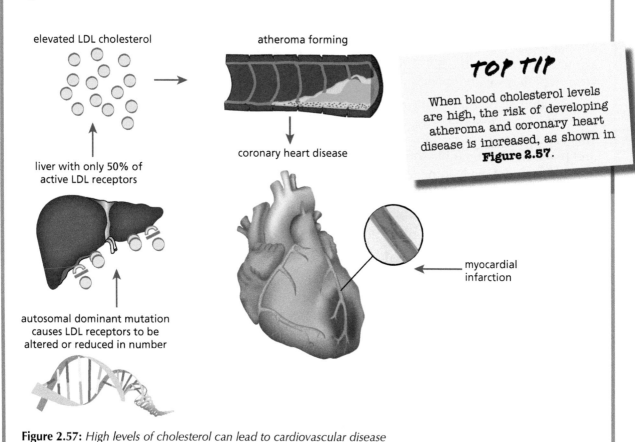

elevated LDL cholesterol

atheroma forming

TOP TIP

When blood cholesterol levels are high, the risk of developing atheroma and coronary heart disease is increased, as shown in **Figure 2.57**.

coronary heart disease

liver with only 50% of active LDL receptors

autosomal dominant mutation causes LDL receptors to be altered or reduced in number

myocardial infarction

Figure 2.57: *High levels of cholesterol can lead to cardiovascular disease*

Blood glucose levels and obesity

Blood glucose levels

Prolonged high levels of blood glucose are now known to provoke atherosclerosis by damaging the endothelial lining of blood vessels and causing fatty streak formations. Elevated glucose levels cause phenotypic changes in the working of the endothelial cells, which absorb more glucose than normal so that the layer becomes thickened and sticky, leading to the formation of atheroma. Atherosclerosis may develop leading to cardiovascular disease, stroke or peripheral vascular disease (PVD). In PVD small blood vessels such as arterioles are particularly prone to damage by above normal glucose levels. This is commonly seen in patients with diabetes.

In untreated diabetes, the endothelial cells of small blood vessels absorb more glucose. This can lead to damage such as bursting and bleeding of retinal blood vessels in the eye, kidney damage leading to renal failure or peripheral nerve dysfunction.

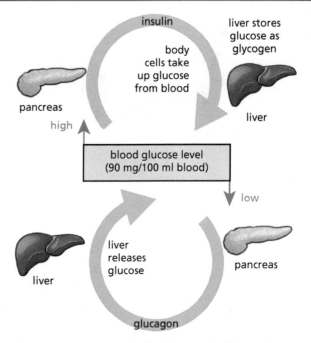

Figure 2.58: *Negative feedback control of blood glucose*

The regulation of blood sugar levels is very carefully controlled within narrow limits using negative feedback (**Figure 2.58**).

Negative feedback control

Following a meal, blood glucose levels become raised. Receptors in the pancreas respond by increasing secretion of the hormone insulin. Insulin is transported in the blood to liver cells where it binds to receptors and causes an increase in glucose uptake. Insulin activates the conversion of glucose to glycogen resulting in a decrease in circulating blood glucose back to set point.

In the evening, following exercise or during sleep, pancreatic receptors respond to lowered blood glucose levels by increasing secretion of glucagon. This hormone activates the conversion of glycogen to glucose in liver cells. The result is an increase in blood glucose concentration back to set point levels.

Sometimes in crisis, exercise or 'fight or flight' responses, glucose concentrations in the blood are raised. This is achieved by the action of adrenaline, released from the **adrenal glands** which are located above the kidneys. Adrenaline stimulates glucagon secretion and inhibits insulin secretion thus increasing circulating blood glucose levels needed to produce additional energy quickly.

In a suddenly stressful situation, glucose is required very quickly to power the fight or flight response (**Figure 2.59**).

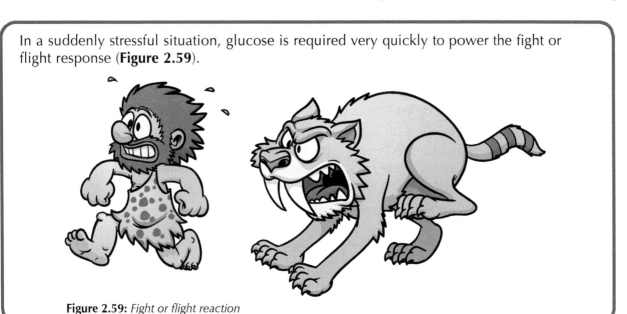

Figure 2.59: *Fight or flight reaction*

Type 1 and type 2 diabetes

There are two forms of diabetes:

1. Type 1 diabetes usually develops in young people. They have normal levels of insulin receptors on their cells but the pancreas produces very little or no insulin at all. This type of diabetes is treated with regular doses of insulin via injection or an insulin pump.

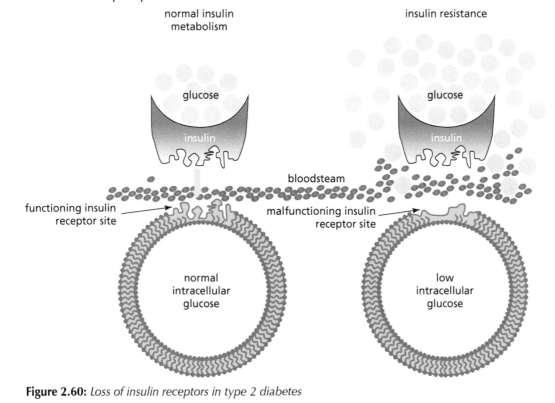

Figure 2.60: *Loss of insulin receptors in type 2 diabetes*

2. Type 2 diabetes typically develops in adulthood. The individual can produce insulin but their cells are less sensitive to it due to insulin resistance. Resistance is caused by a decrease in the number of insulin receptors, as shown in **Figure 2.60**. Thus normal conversion of glucose to glycogen in liver cells does not occur. The likelihood of developing type 2 diabetes is increased by being overweight.

When untreated, individuals with both types of diabetes will show a marked increase in blood glucose concentrations following a meal. The kidneys cannot reabsorb the high levels of glucose from their filtrate back into the bloodstream and as a result glucose appears in the urine. Testing urine for glucose is often used as an indicator of diabetes.

Glucose tolerance test

The ability of the body to respond to ingested glucose is known as its glucose tolerance. A glucose tolerance test aims to identify diabetics as their response to increased blood glucose is impaired.

Initially, blood glucose concentrations are measured after fasting for 8 hours. The individual then drinks a known mass of glucose in solution. Over a minimum of 2 hours following glucose ingestion alterations in their blood glucose concentration are measured. As can be seen in **Figure 2.61** the blood glucose concentration of a diabetic starts at and increases to a much higher level than that of a non-diabetic. Blood glucose concentration takes much longer to return to its starting concentration in a diabetic compared to a non-diabetic.

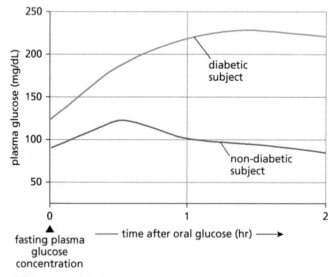

Figure 2.61: *Results of a glucose tolerance test*

Obesity in adults and children is of concern because it is linked to high blood pressure, type 2 diabetes, some forms of cancer and cardiovascular disease (**Figure 2.62**).

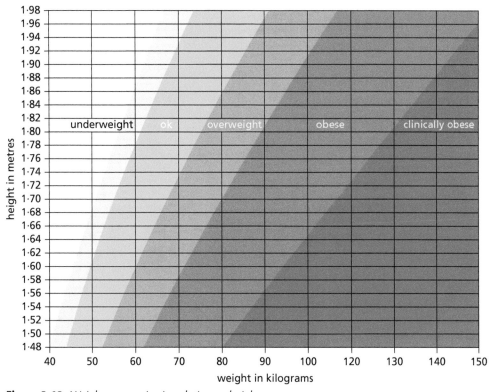

Figure 2.62: *Weight categories in relation to height measurements*

Obesity is prevalent in affluent societies. It is a condition where there is too much body fat in relation to lean body tissue such as muscle. Obesity may impair health as it is a major risk factor for cardiovascular disease and type 2 diabetes.

One of several ways of classifying obesity is the **body mass index (BMI)**, which is calculated:

$$\text{body mass (kg)} / [\text{height (m)}]^2$$

For adults a BMI score of 18·5–24·9 indicates a healthy weight. A BMI score of 25–29·9 indicates an individual is overweight. A BMI score of 30 or higher is classified as obese. It should be noted that sometimes BMI can wrongly classify muscular individuals as obese.

Although some genetic and hormonal causes have been identified, generally obesity is caused by energy intake exceeding net energy expenditure. Contributing factors to the development of obesity include lower levels of physical activity and diets with higher levels of fat than normal. Fat intake should be limited in order to reduce the risk of developing both CVD and obesity. Careful control of free sugar intake is also imperative.

Regular exercise is key to good health as it increases energy expenditure and preserves lean body tissue. Exercise can also help to reduce risk factors for developing CVD by improving blood lipid profiles, reducing stress and high blood pressure and keeping weight under control, as shown in **Figure 2.63**.

> **TOP TIP**
>
> Free sugars can be added in any form; all sugars naturally present in fruit and vegetable juices, purees and pastes and similar products in which the structure has been broken down; all sugars in drinks (except for dairy-based milks); and lactose and galactose added as ingredients.

Figure 2.63: *The obesity solution*

> **TOP TIP**
>
> Individuals are often advised to join a group to enhance their chances of persisting with a diet or an exercise programme.

> **TOP TIP**
>
> Remember that digestion of ingested free sugars requires no metabolic energy. Fats have a high calorific value per gram in comparison to other food groups and should be eaten as part of a balanced diet.

 GOT IT? ☐ ☐ ☐

Area 2 Revision Test

Key area 1 – Gamete production and fertilisation

1. The diagram below represents the appearance of an ovary when viewed under a microscope.

    (a) Name structure X. (1)

    (b) Which number shown in the diagram points to the most mature follicle. (1)

    (c) Apart from forming and releasing ova, state **two** other important functions carried out by the ovaries. (2)

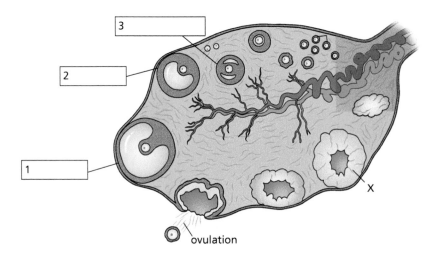

Key area 2 – Hormonal control of reproduction

2. The diagram below illustrates some of the events involved in sperm production.

   (a) State the type of hormone which stimulates the pituitary and is released by the hypothalamus. (1)

   (b) Name the hormone secreted by the pituitary which promotes the secretion of testosterone. (1)

   (c) (i) Apart from the hormone named in (b), state the other hormone secreted by the pituitary which affects the process shown. (1)

   (ii) State the function of this hormone. (1)

   (d) Explain how this process is an example of negative feedback. (2)

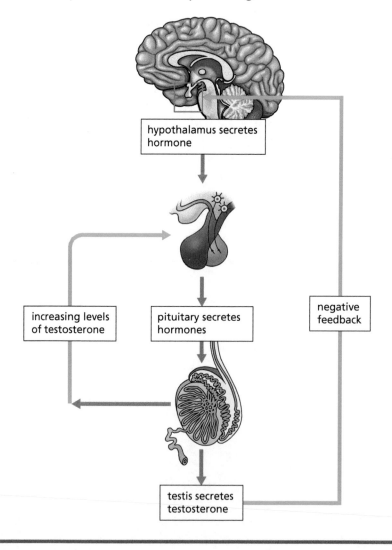

Key area 3 – The biology of controlling fertility

3. (a) Decide if each of the following statements about contraception is **True** or **False** and tick (✔) the correct box.

   If the answer is **False**, write the correct word(s) in the **Correction** box to replace the word(s) <u>underlined</u> in the statement. (3)

Statement	True	False	Correction
<u>contraception</u> is the deliberate prevention of conception by natural or artificial means			
the progesterone-only (mini) pill is an example of a <u>barrier</u> method of contraception			
vasectomy is usually <u>irreversible</u>			

   (b) The table below refers to semen samples in single ejaculations taken from four men.

Semen sample	A	B	C	D
number of sperm in single ejaculation (million)	56	64	50	54
volume of semen (cm³)	1·7	3·2	2·9	1·8

   (i) Identify which man has the highest number of sperm/cm³. (1)

   (ii) The World Health Organization defines a normal sperm count as at least 20 million/cm³.

   Identify which individual does not have a normal sperm count. (1)

(c) The graph shows the blood testosterone levels, measured in ng/100 cm³, in young males aged between 12 and 17 years.

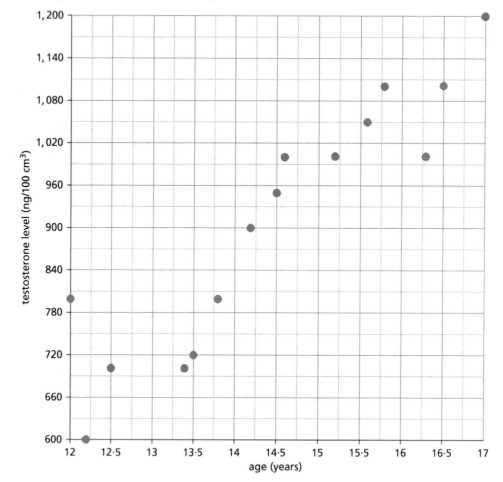

State the range of testosterone levels (ng/100 cm³) for these young males. (1)

Key area 4 – Antenatal and postnatal screening

4. (a) Give **one** useful piece of information about a developing fetus which can be obtained from antenatal screening. (1)

   (b) Ultrasound imaging is a technique used to create an image of a developing baby.

   State **one** use of an anomaly scan. (1)

   (c) State **one** advantage of chorionic villus sampling as a diagnostic tool. (1)

5. The following family tree shows the inheritance of a sex-linked trait caused by a recessive allele (a).

   (a) Using conventional symbols, identify the genotype of individual w. (1)

   (b) Explain why two of the sons in the F_3 (third) generation are affected but one is not. (1)

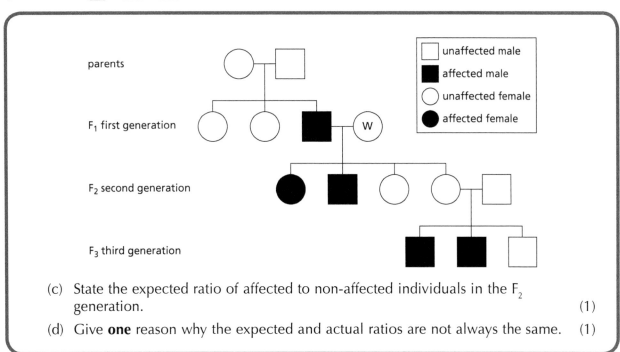

(c) State the expected ratio of affected to non-affected individuals in the F_2 generation. (1)

(d) Give **one** reason why the expected and actual ratios are not always the same. (1)

Key area 5 – The structure and function of arteries, capillaries and veins

6. The diagram below shows some aspects of the circulatory system.

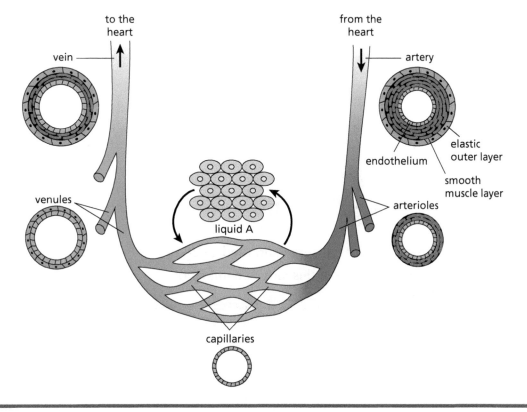

(a) State the name of the structures which are found in the outer layer of blood vessels and help give them the ability to expand and contract. (1)

(b) State the name of liquid A. (1)

(c) Explain how the structure of the endothelium of capillaries makes them efficient for the exchange of materials. (1)

(d) State **one** way in which the central lumen of an artery generally differs from that of a vein. (1)

(e) State **one** difference between tissue fluid and blood plasma. (1)

Key area 6 – The structure and function of the heart

7. The following diagram shows some of the structures associated with the cardiac cycle.

   (a) (i) Identify which of the two structures is responsible for originating the heartbeat. (1)

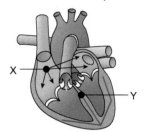

   (ii) Name the structure you have identified. (1)

   (iii) State the term used to describe the special property of cells found in the sino-atrial node. (1)

   (b) State in which phase of the heartbeat the left atrio-ventricular valve will be open. (1)

Key area 7 – Pathology of cardiovascular disease (CVD)

8. The diagram below shows a change which has taken place in an artery over a period of time in a patient whose family has a history of cardiovascular disease.

   (a) State the name of the fatty accumulation indicated by the letter A. (1)

   (b) Explain how the formation of this fatty accumulation can lead to an increase in blood pressure. (1)

Key area 8 – Blood glucose levels and obesity

9. The diagram below shows how glucagon and insulin regulate blood glucose levels.

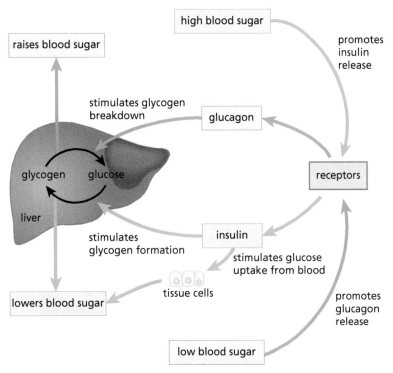

(a) (i) State where the receptors shown on the diagram are located in the body. (1)

(ii) Using the information in the diagram, explain how this illustrates a negative feedback mechanism. (2)

(b) Obesity is linked to cardiovascular disease. One way of measuring obesity is using the body mass index (BMI).

Explain how this can be used to identify individuals who are obese. (2)

(c) A group of slightly overweight men of different ages were tested for the prevalence of high blood pressure. The results are shown in the histogram below.

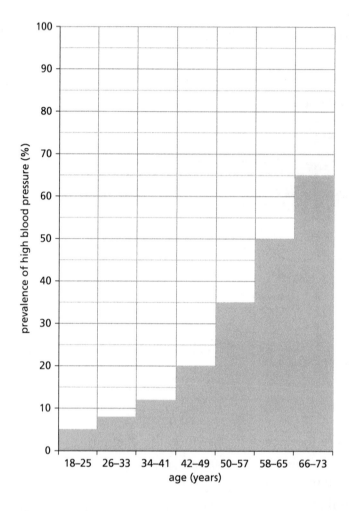

(i) Calculate the ratio of the percentage prevalences of men aged 45 years compared with men aged 62 years expressed as a simple whole number ratio. (1)

(ii) Predict the likely prevalence for men aged 75 years. (1)

Divisions of the nervous system and neural pathways

Central nervous system

The central nervous system (CNS), as shown in **Figure 3.1**, consists of the brain and the spinal cord and is responsible for receiving and sending information, and forming connections with other parts of the nervous system.

Impulses travel from sense organs to the central nervous system via sensory nerves. Motor nerves carry impulses from the central nervous system to muscles and glands.

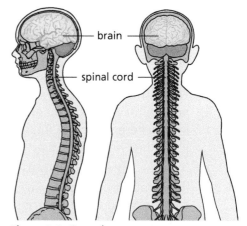

Figure 3.1: *Central nervous system*

Peripheral nervous system

The **peripheral nervous system (PNS)** connects internal or external stimuli with the central nervous system, allowing the body to respond. Two systems make up the peripheral nervous system:

1. the **autonomic nervous system (ANS)**, which consists of the **sympathetic nervous system** and the **parasympathetic nervous system**. It controls and regulates actions which are involuntary. These include breathing, digestion, heartbeat and urinary functions.

2. the **somatic nervous system (SNS)**, which contains sensory and motor neurones, and controls and regulates voluntary and reflex actions. Many of our unconscious reflexes are controlled by the SNS.

The two branches of the autonomic nervous system, sympathetic and parasympathetic, act antagonistically, that is they carry out opposing actions on the same structures, ensuring the optimum conditions for body functions. The sympathetic system speeds up the heart and breathing rates while slowing down peristalsis and the production of intestinal secretions. The parasympathetic system changes these in the opposite way.

TOP TIP

The CNS consists of the brain and spinal cord. The PNS consists of the somatic nervous system (SNS) and the autonomic nervous system (ANS).

TOP TIP

Antagonism is found in many different contexts. For example, the opposite actions of the triceps and biceps.

Neural pathways

Neurons form **synapses** with other neurons in different combinations in the central nervous system. There are three types of pathway:

1. A **convergent neural pathway** where the impulses from two or more neurons travel to one single neuron (**Figure 3.2**).

   The summation of many weak stimuli cause a nerve impulse to be generated in the single neuron on which many other neurons synapse and so increases the sensitivity to inhibitory or excitatory signals. For example, a single motor neuron may need to receive impulses from different parts of the brain before it will fire.

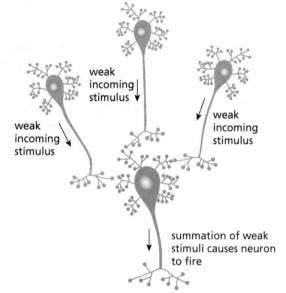

weak incoming stimulus

weak incoming stimulus

weak incoming stimulus

summation of weak stimuli causes neuron to fire

Figure 3.2: *Convergent pathway*

2. A **divergent neural pathway** where the stimuli from one neuron can cause several other neurons to fire in more than one destination. (**Figure 3.3**).

   An impulse from one single neuron generates an impulse which causes two or more neurons to fire, possibly in different parts of the body. For example, an impulse from the hypothalamus can diverge into neurons affecting sweat gland secretion, skeletal muscles and changes in the diameter of blood vessels.

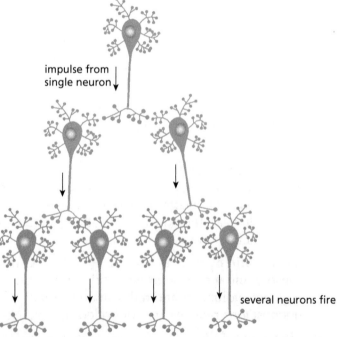

impulse from single neuron

several neurons fire

Figure 3.3: *Divergent pathway*

3. A **reverberating neural pathway** in which an incoming impulse travels along a chain of neurons and back through the pathway ensuring repeated stimulation. (**Figure 3.4**).

The travel of the impulse is maintained by impulses generated by the neurons in the chain, each one of which is linked with the previous cell by further synapses, so allowing repeated stimulation of the pathway. These neurons are involved in the control of activities which are rhythmic, for example, breathing and the sleep–wake cycle.

Previously it was thought that the brain's neural pathways were fixed but it is now known that the brain is in a state of continual change and reorganisation producing new connections, synapses and pathways. This enables humans to acquire new knowledge or learn new skills through experience.

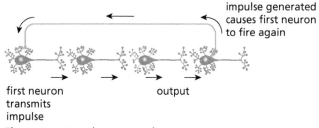

impulse generated causes first neuron to fire again

first neuron transmits impulse

output

Figure 3.4: *Reverberating pathway*

The cerebral cortex

The cerebral cortex

The cerebrum is the largest part of the brain. Its outer layer, the **cerebral cortex** (**Figure 3.5**), receives sensory information, co-ordinates voluntary movement, recalls memories and makes decisions based on experience.

TOP TIP
The cerebral cortex is where conscious thought takes place.

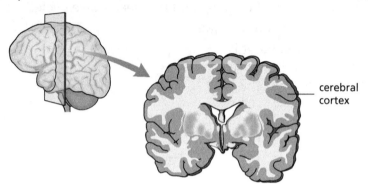

cerebral cortex

Figure 3.5: *The cerebral cortex*

Localisation of brain function

Particular parts of the brain, known as **association areas**, are linked to specific functions as shown in **Figure 3.6**. Association areas are involved in personality, language processing, intelligence and imagination.

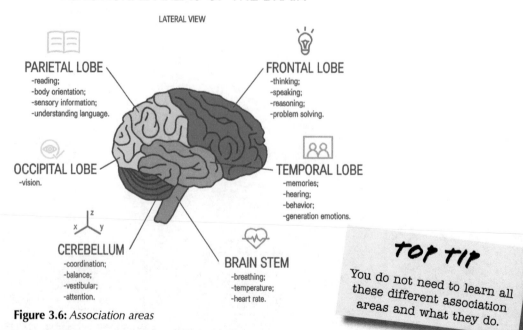

FUNCTIONAL AREAS OF THE BRAIN

LATERAL VIEW

PARIETAL LOBE
-reading;
-body orientation;
-sensory information;
-understanding language.

FRONTAL LOBE
-thinking;
-speaking;
-reasoning;
-problem solving.

OCCIPITAL LOBE
-vision.

TEMPORAL LOBE
-memories;
-hearing;
-behavior;
-generation emotions.

CEREBELLUM
-coordination;
-balance;
-vestibular;
-attention.

BRAIN STEM
-breathing;
-temperature;
-heart rate.

TOP TIP
You do not need to learn all these different association areas and what they do.

Figure 3.6: *Association areas*

The **motor cortex** controls voluntary movements involving muscles of the skeleton.

The **sensory cortex** receives impulses from sense organs such as the skin, muscles and other organs.

The visual association area receives impulses from the eyes and processes these into meaningful images.

Information from one side of the body is received and processed in the opposite half of the cerebrum.

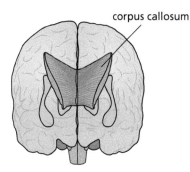

Information from one side of the brain travels to the opposite side via a tract of nerve tissue called the **corpus callosum**, shown in **Figure 3.7**. The left cerebral

Figure 3.7: *The corpus callosum connects both halves of the cerebrum*

hemisphere processes information from the right visual field and controls the right side of the body. The right cerebral hemisphere processes information from the left visual field and controls the left side of the body.

Memory

Memory

Most human behaviour is a function of memory, the ability to retain information about events, stimuli, experiences

Figure 3.8: *Stages of memory*

and ideas, even once the original stimuli are no longer in operation. Memories include past experiences, thoughts and knowledge.

Memory involves three important and linked processes: **encoding**, **storage** and **retrieval** of information (**Figure 3.8**). Encoding processes new facts and other types of information into a form which can be stored and recalled later. Storage maintains information over time. Retrieval is the ability to access information when it is needed.

Detection of environmental stimuli, including touch, sound and light, enters the **sensory memory** but is very short-lived, lasting for only a few seconds. This information is transferred to the **short-term memory (STM)**, which is a temporary storage for information before it has been processed or interpreted. It has a limited capacity of around seven discrete items of information, which are held for a limited time. If the information is of no value, it is discarded, but if it is potentially valuable, it is transferred to a long-term storage called the **long-term memory (LTM)**, as shown in **Figure 3.9**.

> **TOP TIP**
>
> Only selected stimuli, such as sounds and images, are encoded into the short-term memory.

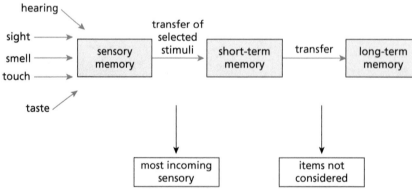

Figure 3.9: *Development of long-term memory*

Short-term memory has a limited span, linked to its limited capacity for storage. It can store information temporarily, enough to carry out relatively complex tasks such as learning and reasoning.

> **TOP TIP**
>
> The average short-term memory span for an adult is about seven numbers. The duration of short-term memory is about 15–30 seconds.

When a number of items, such as a list of words, enter the short-term memory, those which arrive first and those which arrive last are better remembered than those items which arrive in between. This is known as the **serial position effect**, shown in **Figure 3.10**.

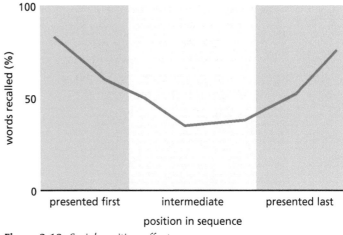

Figure 3.10: *Serial position effect*

One way of enhancing the short-term memory function is by repeatedly going over the items presented, extending the timespan of the information retention until the items become stored in the long-term memory. This is called **rehearsal** and is a common strategy used to retain information such as birthdays or telephone numbers (**Figure 3.11**).

Another commonly used technique for remembering lists of items is to group them in a meaningful way. For example, a candidate's examination number may consist of more than seven numbers but it can be stored as a single item like a birthday or a car registration number. This is called **chunking** (**Figure 3.12**).

The long-term memory has an unlimited capacity for the storage of information. Transferring information from short-term to long-term memory is enhanced by rehearsal, organisation and **elaboration**, or by adding more detail to a memory item making it easier to store by associating it with previously recorded material. The ability to recall anything in long-term memory is known as retrieval (**Figure 3.13**).

Incoming information from the sensory inputs must be encoded to change it into a form which can be stored and recalled later. For example, this could be using mental images, how something felt or sounded, or by forming linkages with information remembered already. Encoding using rehearsal is an example of a shallow type of encoding, while creating elaborative links is an example of a deeper type of encoding.

TOP TIP

Recalling the most recent items depends on short-term memory while recalling the primary items depends on long-term memory.

Figure 3.11: *Enhancing short-term memory function by rehearsal*

Recalling information from the long-term memory is helped if the context of the encoding and retrieval are the same. For example, speaking to a friend might also be linked to the clothes he/she was wearing, the weather or the time of day. A **contextual cue** makes it easier to retrieve a memory when the circumstances around that memory are the same for both the original encoding and the retrieval.

which is easier to memorise?

3,129,823,812

or

312-982-3812

Figure 3.12: *Enhancing short-term memory function by chunking*

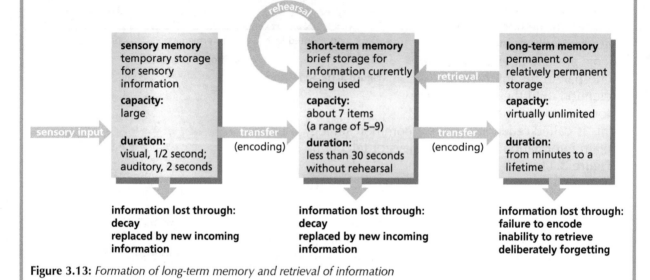

Figure 3.13: *Formation of long-term memory and retrieval of information*

The cells of the nervous system and neurotransmitters at synapses

Neurons

The nervous system is made up of neurons, as shown in **Figure 3.14**, which are capable of rapidly carrying information in the form of nerve impulses.

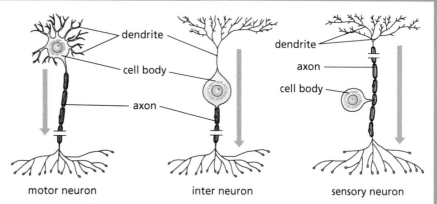

Figure 3.14: *Different types of neurons*

The three types of neuron are sensory, motor and inter. Each has:

- a **cell body** where the nucleus is located
- fibres called **axons** which carry nerve impulses away from the cell body
- fibres called **dendrites** which carry nerve impulses towards the cell body.

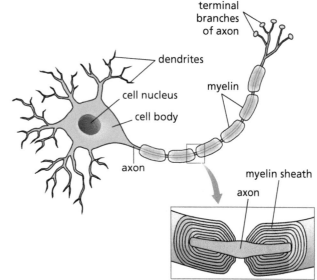

As shown in **Figure 3.15**, long neurons have a surrounding sheath made mainly of a fatty material called **myelin** which insulates the internal nerve fibre and increases the speed of impulse conduction.

Figure 3.15: *Surrounding myelin sheath found in many neurons*

The laying down of myelin starts around the fourteenth week in the development of a fetus, but it is not complete at birth and continues into adolescence. This is why responses in the first two years of life are neither co-ordinated nor rapid (**Figure 3.16**).

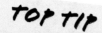

TOP TIP

Certain diseases can destroy the myelin sheath which surrounds and supports neurons. This can lead to a loss of co-ordination.

Figure 3.16: *Development of walking*

Neurons have associated **glial cells** **(Figure 3.17)**, which:

- form the myelin needed for the surrounding sheath
- provide mechanical support by holding the neurons in place
- supply nutrients and oxygen to neurons.

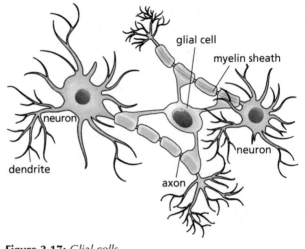

Figure 3.17: *Glial cells*

TOP TIP

In multiple sclerosis, T lymphocytes attack antigens on the myelin sheath of neurons.

Neurotransmitters at synapses

The transmission of impulses from one nerve ending to another requires the presence of a chemical called a **neurotransmitter**. It is released at the end of a nerve fibre and passes across the **synaptic cleft**, shown in **Figure 3.18**, to the next fibre in line, causing it to fire, allowing the impulse to continue.

The neurotransmitter is released from small sacs called **vesicles** in the presynaptic neuron and diffuses towards receptors on the postsynaptic neuron.

Once a neurotransmitter has carried out its function, any still remaining in the synaptic cleft is either degraded enzymatically to non-active products or reabsorbed by the membrane at the end of the fibre along which the impulse was first initiated. This prevents continuous stimulation of the postsynaptic neuron. One common neurotransmitter which is broken down enzymatically is **acetylcholine**.

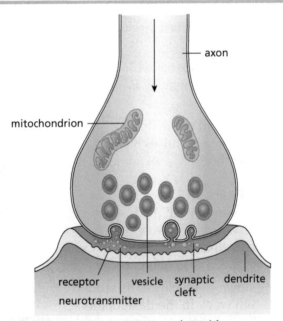

Figure 3.18: *Neurotransmitters released from presynaptic neuron travel across synaptic clefts*

TOP TIP

Neurotransmitters relay impulses across the synaptic cleft. Vesicles also store neurotransmitters.

The transmission of a nerve impulse across a synapse depends on the total secretion of neurotransmitter. If the stimulus is weak, not enough neurotransmitter will be secreted and the next fibre will not fire. Therefore a synapse acts as a filter preventing weak stimuli, such as a quiet sound or dim light, from generating a response. However, if enough weak stimuli are sent to a particular neuron, their individual inputs can collectively cause enough neurotransmitter to be secreted and the next fibre to fire. This is called **summation** (**Figure 3.19**).

A synapse also acts as a mechanism for a neuron to process a variety of incoming stimuli and to decide whether or not to secrete enough neurotransmitter to cause the next fibre to fire. This is because the plasma membrane of the next fibre has two different types of receptors, one promoting and one inhibiting the chance of an impulse being successful. These synapses are called **excitatory** and **inhibitory**, respectively (**Figure 3.20**).

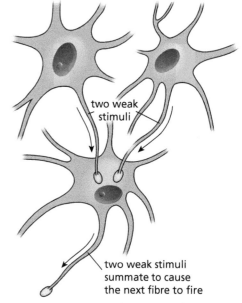

Figure 3.19: *Summation of nerve impulses*

Depending on the overall balance of the cumulative effect of all the excitatory and inhibitory inputs, the impulse either will or will not be transmitted.

Converging neural pathways release enough neurotransmitters to reach the threshold to trigger an impulse.

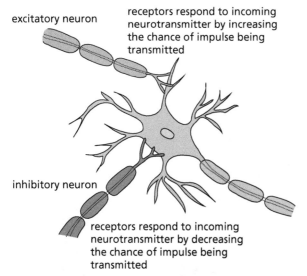

Figure 3.20: *Nerve impulse transmission can be modified by excitatory or inhibitory neuron action*

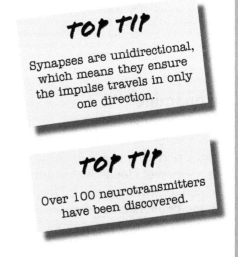

TOP TIP

Synapses are unidirectional, which means they ensure the impulse travels in only one direction.

TOP TIP

Over 100 neurotransmitters have been discovered.

Neurotransmitter effects on mood and behaviour

When the levels of neurotransmitters (**Figure 3.21 and 3.22**) are unbalanced, mood and behaviour can be affected. For example, loss of interest, low motivational drive, erratic behaviour and mood swings are all potentially influenced by neurotransmitter imbalance. Two neurotransmitters which influence mood and behaviour are **endorphin** and **dopamine**.

Endorphin

Endorphin functions in a similar way to naturally occurring painkillers, not only reducing pain but also enhancing the feeling of wellbeing. For example, the reduction of labour pains in childbirth relies on the body's production of endorphin. People with low levels of endorphin tend to feel more anxious and their pain thresholds are lower. Appetite control and the secretion of sex hormones are both linked to endorphin levels. Physical exercise produces a high which is linked to enhanced levels of endorphin.

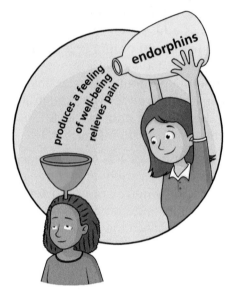

Figure 3.21: *Neurotransmitter endorphin*

> **TOP TIP**
> The body produces about 20 different types of endorphin which can increase the feeling of pleasure from eating and sex.

Dopamine

Dopamine is an inhibitory neurotransmitter. Its presence will block nerve impulse transmission at a synapse. Dopamine is strongly linked to activating reward pathways in the brain, helping to raise motivation levels and promoting a feeling of wellbeing.

> **TOP TIP**
> Low levels of dopamine are observed in patients with uncontrollable muscle contractions, for example, Parkinson's disease.

Figure 3.22: *Neurotransmitter dopamine*

Neurotransmitter-related disorders and their treatment

Agonistic and antagonistic actions

An imbalance of neurotransmitters can result in mental disorders, some of which can be treated using prescribed drugs like **agonists** and **antagonists** (**Figure 3.23**). They act on the receptors in the membranes of neurons, by either enhancing or inhibiting nerve impulse transmission.

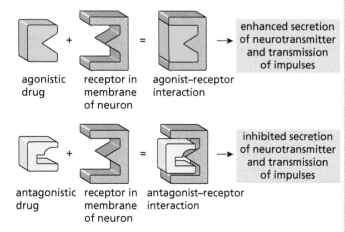

agonistic drug + receptor in membrane of neuron = agonist–receptor interaction → enhanced secretion of neurotransmitter and transmission of impulses

antagonistic drug + receptor in membrane of neuron = antagonist–receptor interaction → inhibited secretion of neurotransmitter and transmission of impulses

Figure 3.23: *Agonistic and antagonistic actions*

> ### TOP TIP
> Agonists mimic the action of neurotransmitters at the synapse while antagonists bind to specific receptors, blocking synaptic transmission.

Certain drugs can act as inhibitors of the normal enzyme function which breaks down a neurotransmitter (**Figure 3.24**).

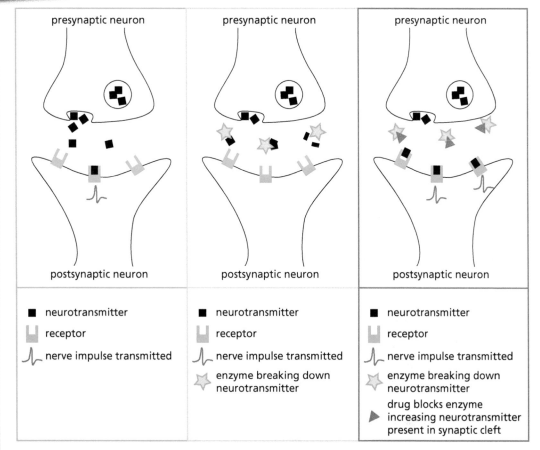

- ■ neurotransmitter
- receptor
- nerve impulse transmitted

- ■ neurotransmitter
- receptor
- nerve impulse transmitted
- enzyme breaking down neurotransmitter

- ■ neurotransmitter
- receptor
- nerve impulse transmitted
- enzyme breaking down neurotransmitter
- ▶ drug blocks enzyme increasing neurotransmitter present in synaptic cleft

Figure 3.24: *Action of drugs on enzymes involved in neurotransmitter function*

Other drugs slow down the reabsorption of the neurotransmitter after an impulse has crossed the synaptic cleft (**Figure 3.25**).

Many drugs used to treat the imbalances of neurotransmitters are themselves very similar in their chemical structure to those neurotransmitters.

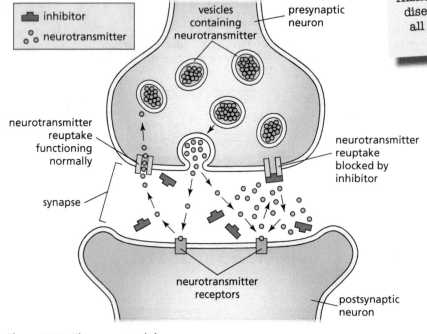

Figure 3.25: *The synaptic cleft*

Mode of action of recreational drugs

Recreational drugs, as shown in **Figure 3.26**, are chemicals which mimic neurotransmitters by acting as agonists or antagonists and are taken for non-medical reasons such as:

- peer pressure
- experimentation with new experiences
- increased feelings of euphoria, relaxation and self-confidence
- relief of anxiety or stress
- performance enhancement.

By affecting the reward neural pathways in the brain these drugs can alter mood, perception, behaviour and cognition.

Figure 3.26: *Various chemicals used as recreational drugs*

Some recreational drugs such as nicotine, alcohol and caffeine are legal.

Other recreational drugs such as ecstasy, cocaine and cannabis are illegal.

Repeated exposure to a drug can alter its effect in different ways, shown in **Figure 3.27**.

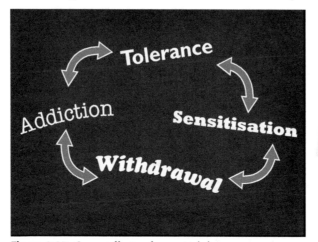

Figure 3.27: *Some effects of repeated drug exposure*

> **TOP TIP**
>
> Recreational drugs affect neurotransmission at synapses in the brain, altering an individual's mood, cognition, perception and behaviour.

Sensitisation, shown in **Figure 3.28**, results from prolonged exposure to an antagonistic drug. There is an increase in the number and sensitivity of the binding neurotransmitter receptors formed. This may result in addiction and the experience of severe withdrawal symptoms if the drug is no longer taken so the individual craves more of the drug.

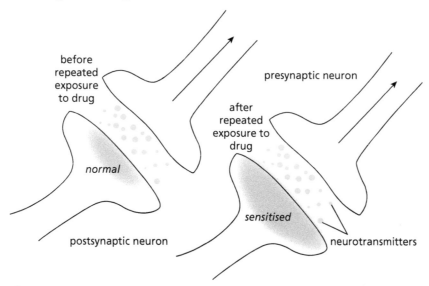

Figure 3.28: *Sensitisation to a drug after repeated exposure*

Desensitisation results from prolonged exposure to an agonistic drug. There is a decrease in the number and sensitivity of the binding neurotransmitter receptors formed. This may result in the same dose of the drug having less effect so that an increased dosage is required to achieve a similar high. Drug **tolerance** develops as this process continues.

Non-specific body defences

Physical and chemical defences

Epithelial cells line the cavities and surfaces of blood vessels and organs throughout the body. They form a physical barrier and produce chemical **secretions** against **pathogens**. Some examples of these are shown in **Figure 3.29**. The skin, respiratory and digestive systems all contain closely packed epithelial cells.

During an **inflammatory response** there is a release of **histamine** by **mast cells** causing vasodilation and increased capillary **permeability**. The increased blood flow and secretion of **cytokines** leads to an accumulation of **phagocytes** and the delivery of clotting elements to the site of infection. This is shown in **Figure 3.30**.

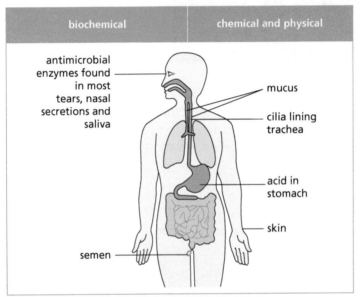

TOP TIP

A pathogen is any organism, such as a bacterium or virus, that can cause disease. Chemical secretions are produced by the body cells against invading pathogens.

Figure 3.29: *Examples of secretions*

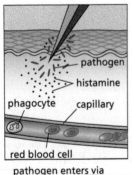

pathogen enters via wound

release of histamine by mast cells

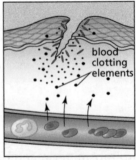

histamine causes vasodilation and increased capillary permeability

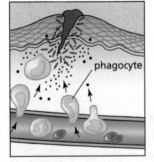

increased blood flow and secretion of cytokines

accumulation of phagocytes and the delivery of clotting elements

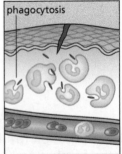

wound clotted

pathogen destroyed by phagocytosis

Figure 3.30: *Inflammatory response*

The role of a phagocyte is to detect pathogens and destroy them by a process called **phagocytosis**, as shown in **Figure 3.31**. Phagocytes detect chemicals given off by a pathogen and move towards it. The pathogen is engulfed by phagocytosis. It is broken down by digestive enzymes released from phagocyte organelles called lysosomes. Simultaneously the phagocyte releases protein

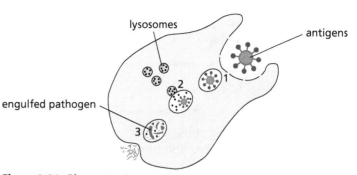

Figure 3.31: *Phagocytosis*

molecules called cytokines which attract more phagocytes to the site of infection, as shown in **Figure 3.32**.

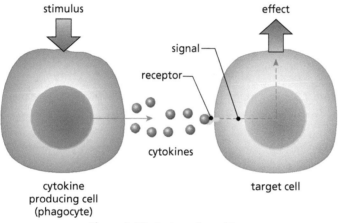

Figure 3.32: *Action of cytokines*

Specific cellular defences against pathogens

Lymphocytes

A range of white blood cells, two of which are shown in **Figure 3.33**, are continually circulating and monitoring the tissues. If tissues become damaged or are invaded, the cells release cytokines, increasing blood flow which results in white blood cells accumulating at the site of infection or tissue damage.

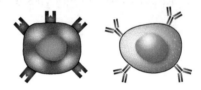

T lymphocyte   B lymphocyte

Figure 3.33: *Two examples of white blood cells involved in defence*

Lymphocytes and antigens

Lymphocytes are specialised cells derived from bone marrow tissue stem cells. They are white blood cells and are involved in the **specific immune response**. Lymphocytes respond to invading pathogens by binding their specific membrane receptors to specific **antigens** found on the pathogens. When the antigen and receptor bind the lymphocyte divides repeatedly resulting in the formation of a clonal population of identical lymphocytes, as shown in **Figure 3.34**.

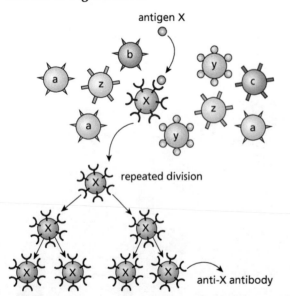

TOP TIP

Antigens are molecules, often proteins, located on the surface of cells, that trigger a specific immune response.

Figure 3.34: *Antigen binding results in a clonal population*

T and B lymphocytes

Lymphocytes can be divided into two types: **B lymphocytes** and **T lymphocytes**.

T lymphocytes recognise antigens originating from pathogens on the membranes of infected cells.

T lymphocytes attach onto infected cells and release proteins. These proteins diffuse into the cells causing production of self-destructive enzymes which destroy the cells. This process is known as **apoptosis** or programmed cell death.

The remains of the cell are then removed by phagocytosis.

T lymphocytes can normally recognise the difference between antigens on their own body cells, called self-antigens, and non-self antigens found on the membranes of infected cells.

Sometimes T lymphocytes respond to their own self-antigens. This is a serious failure of the immune system and is known as **autoimmunity** (**Figure 3.35**). In autoimmunity, the T lymphocytes attack the body's own cells. This causes autoimmune diseases such as type 1 diabetes and rheumatoid arthritis, as shown in **Figure 3.36**.

normal immune response

antigens invade

antibodies form

antibodies destroy invading antigens

antibodies remain and protect

autoimmune disease

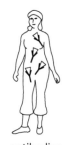
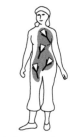

immune system forms antibodies to self-antigens

antibodies attack self-antigens

inflammation and tissue damage

Figure 3.35: *Autoimmunity caused by a failure in the immune system*

TOP TIP

In rheumatoid arthritis, cells in the joints produce cytokines that provoke joint destruction.

Rheumatoid arthritis

X-RAYS

Women are 3 times more likely to develop RA than men

Vaccination

Management

80% 80% of RA patients between the ages of 35 and 50

Risk

Heredity Age Lifestyle Pollution

Complications

Heart attack Stroke

70% of RA patients have wrist and hand problems

90% of RA patients have symptoms in the foot

Exercise Surgery Dietary supplements

Antirheumatic drugs Stop smoking Limit alcohol

Figure 3.36: *Overview of rheumatoid arthritis (RA)*

In type 1 diabetes, T lymphocytes attack insulin-producing cells. People who have this type of diabetes will inject themselves daily with insulin, as shown in **Figure 3.37**.

Figure 3.37: *A person with type 1 diabetes taking one of their regular insulin doses*

Allergy is a hypersensitive B lymphocyte response to an antigen that is normally harmless. Examples include hayfever **(Figure 3.38)**, asthma and food allergies.

Figure 3.38: *Hayfever*

TOP TIP

Hayfever is an allergic reaction where the B lymphocytes overreact to the antigens on pollen grains.

B lymphocytes

B lymphocytes recognise and bind to antigens which may be free or attached to the surface of pathogens. Each B lymphocyte can recognise pathogenic antigens and produce specific antibodies against this antigen. The resultant antigen–antibody complex inactivates the pathogen and results in its destruction by phagocytes.

As shown in **Figure 3.39**, antibodies are Y-shaped proteins that are made in response to antigens. They are secreted into the plasma where they circulate and their specific receptor binding sites recognise a particular antigen on a pathogen.

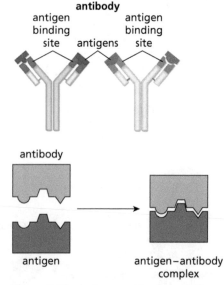

Figure 3.39: *Antigen–antibody complexes*

Immunological memory

Some T and B lymphocytes produced in response to antigens by **clonal selection** survive as **memory cells**. A **secondary exposure** to the same antigen gives rise to a new clone of specific lymphocytes producing a greater secondary immunological response. This is shown in **Figure 3.40**. where antibody production can be seen to be greater and more rapid in the secondary response than during the primary response. In fact, the invading pathogens are destroyed before the person even shows signs of symptoms.

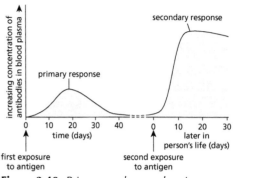

Figure 3.40: *Primary and secondary immune responses*

Sometimes a new emerging disease appears. This was the case for **acquired immune deficiency syndrome** (AIDS), which first appeared in the news in 1981. The syndrome causes a deficiency in the immune system due to infection with **human immunodeficiency virus** (HIV). HIV invades and destroys T lymphocytes. This causes depletion of T lymphocytes, leading to the development of AIDS, as shown in **Figure 3.41**. Individuals with AIDS have a weakened immune system and so are more vulnerable to opportunistic infections and forms of cancer.

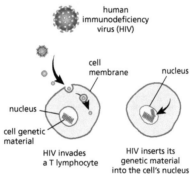

Figure 3.41: *HIV attack on lymphocytes*

> ## TOP TIP
> **Immunodeficiency** is the inability of the body to make a normal immune response to an antigen.

> ## TOP TIP
> A syndrome is a collection of symptoms which occur together.

Immunisation

Vaccination

A vaccine is a suspension of a modified, inactivated or dead pathogen or parts of a pathogen or an inactivated toxin. Vaccination against disease stimulates production of antibodies in response to infectious pathogenic antigens. Immunity is developed by lymphocytes which produce memory cells in response to pathogenic antigens.

Antigens are usually mixed with an adjuvant when producing the vaccine. An adjuvant is an additional agent, such as detergent or oil, that makes the vaccine more effective, thereby increasing the immune response.

Vaccination provides effective control over many common viral and bacterial diseases, as shown in **Figure 3.42**.

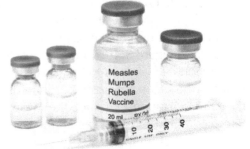

Figure 3.42: *MMR vaccination*

> **TOP TIP**
>
> The measles, mumps and rubella (MMR) vaccine protects a child against three diseases in one injection.

Herd immunity

Disease control does not require everyone in the population to be vaccinated (immunised). In herd immunity most of the population is immune due to immunisation, and outbreaks are limited to small numbers of non-immune individuals. These non-immune individuals are protected as they are less likely to come into contact with infected individuals, as shown in **Figure 3.43**.

Establishing herd immunity is important in reducing the spread of diseases. The herd immunity threshold depends on the type of disease, the effectiveness of the vaccine and the density of the population.

> **TOP TIP**
>
> The herd immunity threshold varies for different pathogens. For example, for measles about 92% of the population need to be vaccinated; for influenza it is 35–75%.

Public health immunisation programmes

Mass vaccination programmes are designed to establish herd immunity to a disease.

In the developing world where poverty and poor nutrition frequently coexist it is often very difficult to introduce a widespread vaccination programme. In the developed world adverse publicity and Internet scaremongering about the safety of vaccines can lead to a decrease in uptake of readily available vaccines. In the United Kingdom recent adverse publicity relating to the measles, mumps and rubella (MMR) vaccine has led to increasing outbreaks of measles in the population.

> **TOP TIP**
>
> In some cases, a vaccination can be rejected by a percentage of the population in the developed world.

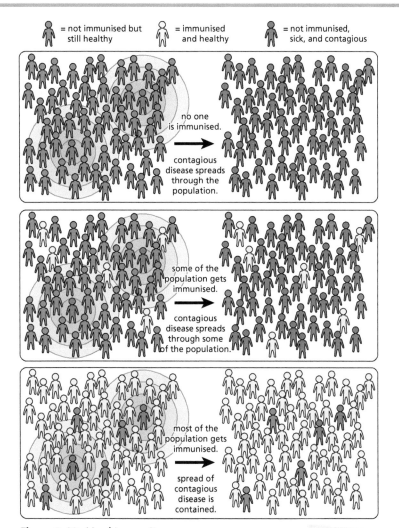

Figure 3.43: *Herd immunity*

TOP TIP

Several pathogens have evolved mechanisms that evade the specific immune system: this has consequences for vaccination strategies.

Antigenic variation

Some pathogens can change their antigens, enabling them to avoid the impact of immunological memory cells and destruction by the immune system. This is called antigenic variation and is shown in **Figure 3.44**.

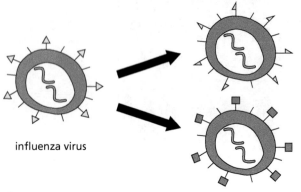

influenza virus

> ***TOP TIP***
>
> Antigenic variation occurs in the influenza virus. This is why it remains a public health problem and why those most at risk are vaccinated every year.

Figure 3.44: *Antigenic variation*

Antigenic variation occurs in infections such as influenza. This is one of the reasons why such infections are still common in many parts of the world. Creating a vaccine for a continually changing pathogen is difficult and the reason why influenza is still a major health problem needing new yearly vaccinations.

Clinical trials of vaccines and drugs

Clinical trials of vaccines and drugs

Before being approved and licensed, vaccines and drugs are subjected to clinical trials to establish their safety and effectiveness, as shown in **Figure 3.45**.

The careful design of clinical trials is vital in order to produce valid results. For a clinical trial to be effective it must have randomised, **double-blind** and **placebo-controlled** protocols. A **placebo**, as shown in **Figure 3.46**, is used to eliminate any effects of a treatment which do not depend on the drug itself.

Clinical trial design factor	Reason
subjects assigned groups randomly	reduces bias in relation to gender and age
one group receives a placebo control the other group receives the vaccine	ensures a valid comparison can be made
subjects or researchers do not know whether the subjects are receiving trial drug or placebo control (double blind)	ensures non-biased interpretation of results
large group size	reduces experimental error necessary for statistical tests

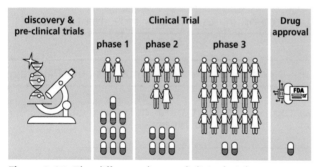

Figure 3.45: *The different phases of clinical trials*

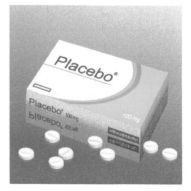

Figure 3.46: *Clinical trials make use of placebos*

Area 3 Revision Test

Key area 1 – Divisions of the nervous system and neural pathways

1. (a) The diagram below shows a side view of some areas of the brain.

   Complete the following table by identifying the regions of the brain labelled A and C and state **one** function of the *corpus callosum*. (3)

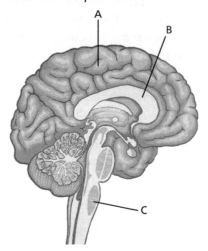

	Region of brain	Example of one function
A		co-ordinates voluntary movements
B	*corpus callosum*	
C		involuntary control of intercostal muscles and diaphragm

   (b) The diagram below shows one way of dividing the central nervous system:

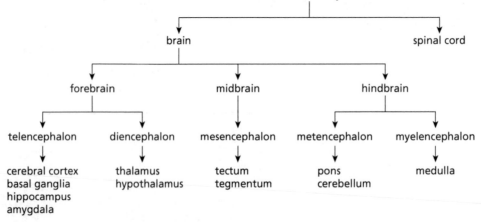

   Using information from the diagram state:

   (i) Where in the brain the tegmentum is located. (1)

   (ii) **Two** facts about where the hippocampus is located in the brain. (2)

Key area 2 – The cerebral cortex

2. (a) The surface of the cerebral cortex is highly folded.

    Suggest **one** possible advantage for this arrangement. (1)

    (b) State the structure which allows each half of the cerebrum to communicate with the other half. (1)

Key area 3 – Memory

3. After perception has taken place, the information may be passed to long-term memory via short-term memory.

    (a) Describe **two** ways in which the transfer of information from short- to long-term memory can be brought about. (2)

    (b) Explain how chunking helps short-term memory function. (1)

    (c) Explain what is meant by a memory 'cue'. (1)

Key area 4 – The cells of the nervous system and neurotransmitters at synapses

4. The diagram below shows a sensory neuron and its link to receptors in the skin:

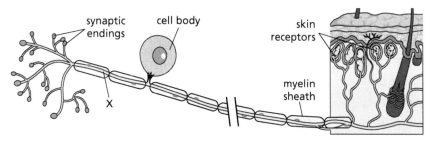

    (a) Name structure X. (1)

    (b) State **one** way in which the myelin sheath helps the transmission of a nerve impulse. (1)

    (c) At the synaptic endings, a neurotransmitter is released and comes into contact with the next neuron in line.

        (i) State how the neurotransmitter is stored in the synaptic endings. (1)

        (ii) Some pain-killing drugs act by binding to the neurotransmitter. Predict what would happen to the perception of pain if one of these drugs was administered. (1)

Key area 5 – Non-specific body defences

5. (a) (i) Name the cells which can form a physical barrier and produce secretions against infection. (1)

   (ii) During an inflammatory response, name the chemical released by mast cells. (1)

   (iii) Describe **one** effect of the release of this chemical. (1)

   (b) Describe how phagocytes protect the body against pathogens. (1)

Key area 6 – Specific cellular defences against pathogens

6. (a) Draw lines to match each of the following terms with its correct description. (3)

Term	Description
T lymphocyte	Surface protein which triggers a specific immune response
B lymphocyte	Induces apoptosis
Antigen	Secretes antibodies

   (b) Describe what leads to autoimmune disease. (1)

7. Decide if each of the following statements related to specific cellular defences is **True** or **False** and tick (✔) the correct box.

   If the answer is **False**, write the correct word(s) in the **Correction** box to replace the word underlined in the statement. (3)

Statement	True	False	Correction
infectious diseases are caused by <u>pathogens</u>			
an allergy is a hypersensitive <u>T lymphocyte</u> response to antigen that is normally harmless			
antibodies are secreted by <u>B lymphocytes</u>			

Key area 7 – Immunisation

8. The table below shows the number of people infected with human immune deficiency virus (HIV) globally and the number with access to treatment.

Year	People living with HIV (millions)	People accessing treatment (millions)
2011	33·9	9·1
2012	34·5	11
2013	35·2	13
2014	35·9	15
2015	36·6	17
2016	36·7	19·5
2017	36·9	21·7
2018	37·9	23·3

   (a) Select the year when the number of people living with HIV was less than 35 million but the people accessing treatment was more than 10 million. (1)

   (b) Calculate the percentage increase in people accessing treatment between 2010 and 2014. (1)

   (c) State **two** conclusions that can be drawn from the data. (2)

   (d) Describe the effect of HIV on the immune system. (1)

9. (a) Describe the meaning of the term herd immunity. (1)

   (b) State **one** of the difficulties with widespread vaccination. (1)

   (c) Describe what is meant by the term antigenic variation. (1)

Key area 8 – Clinical trials of vaccines and drugs

10. (a) State why clinical trials are randomised. (1)

   (b) State how a valid comparison could be ensured in a double-blind trial. (1)

   (c) Explain why low numbers of subjects are not appropriate for clinical trials. (1)

Glossary

accessory gland: group of cells attached to a larger gland which share a similar structure and function

acetyl coenzyme A: important intermediate metabolite linking glycolysis to the citric acid cycle

acetylcholine: chemical neurotransmitter found in many different locations

acquired immune deficiency syndrome (AIDS): disease caused by the immunodeficiency virus which targets the immune system

activation energy: the energy needed to break chemical bonds in the reactant chemicals

active site: area of the enzyme which binds to the substrate

adjuvant: a substance which enhances the body's immune response to an antigen

adrenal gland: endocrine gland which secretes several important hormones including adrenaline

affinity: the force by which atoms are held together in chemical compounds

agonistic drug: chemical which enhances the action of another chemical, such as a neurotransmitter

allergy: a hypersensitive B lymphocyte response to an antigen that is normally harmless

amniocentesis: technique used to diagnose potential congenital abnormalities by examining cells found in the amniotic fluid

amniotic fluid: watery liquid produced from the mother's plasma surrounding and giving protection to the developing fetus

amplification: creating many copies of a fragment of DNA

anabolism: synthesis of new molecules from basic building blocks using energy in the form of ATP

anomaly scan: scan that takes place between 18 and 20 weeks of pregnancy. It may be able to detect serious physical abnormalities in the fetus

antagonism: having the opposite effect to another agent

antagonistic drug: chemical which inhibits the action of another chemical, such as a neurotransmitter

antenatal screening: use of diagnostic tests to look for any possible issues with a pregnancy

antibody: a protein produced by the body's immune system when it detects harmful substances, called antigens

anticodon: triplet of bases in tRNA that codes for a specific amino acid and is complementary to a specific codon in mRNA

antigen: molecule, often a protein, located on the surface of cells, that triggers a specific immune response

antigen–antibody complex: a molecule formed from the binding of an antibody to an antigen

antigenic variation: the mechanism by which a pathogen alters its surface proteins in order to evade a host immune response

antiparallel: parallel, but running in opposite directions

apoptosis: a process of programmed cell death

arteriole: a small division of an artery

artificial insemination: introduction of sperm into the reproductive tract using an artificial means rather than sexual intercourse

association area: part of the brain dedicated to a particular function

atheroma: build-up of fatty substances on the walls of arteries

ATP: adenosine triphosphate. High-energy molecule produced during cellular respiration

ATP synthase: an enzyme that catalyses the synthesis of ATP from ADP and inorganic phosphate

atrio–ventricular node: cluster of cells located in the wall of the right atrium, near the base, which passes the incoming stimulus from the sino-atrial node to the walls of the ventricles

atrio-ventricular valve: structure preventing the backflow of blood from the ventricles into the atria

autoimmunity: the body's immune system attacks and destroys healthy body tissue by mistakenly failing to recognise self-antigens

autonomic nervous system (ANS): nervous system associated with unconscious control of internal body functions

autosome: chromosome other than a sex chromosome

axon: fibre in a neuron which carries nerve impulses away from cell body

barrier: contraceptive method which prevents egg and sperm meeting

bioinformatics: computer technology used to identify DNA sequences

biosynthetic process: biological molecules join together to form larger, more complex substances, such as amino acids joining together to form proteins

B lymphocyte: white blood cell which produces specific antibodies against antigens leading to the destruction of the pathogen

body mass index (BMI): measurement involving height and body weight which can indicate levels of overweight

cancer cell: cell that divides relentlessly, forming solid tumours or flooding the blood with abnormal cells

cardiac cycle: regular contraction and relaxation of the heart chambers which make up one single beat

cardiac output: volume of blood pumped by either the left or right ventricle per minute

catabolism: complex molecules are broken down into smaller subunits with the release of energy

cell body: structure in a neuron where the nucleus is located

cellular differentiation: the process by which an unspecialised cell changes in order to carry out a specific function.

central lumen: space within the blood vessels through which blood travels

cerebral cortex: outer layer of brain

cervical cap: artificial barrier method of contraception which fits over the cervix and prevents sperms reaching ova

cholesterol: insoluble lipid which forms part of an animal cell membrane

chorionic villus sampling: technique used to diagnose potential congenital abnormalities by examining cells from the placenta

chunking: organising distinct bits of information into a single whole, making them easier to recall

citrate: intermediate compound in the citric acid cycle

citric acid cycle: cyclical series of reactions operating under aerobic conditions in the mitochondrial matrix

clonal selection: antigen binding leads to repeated lymphocyte division resulting in a clonal population of lymphocytes

coding: sequence of DNA bases that enters transcription and translation, resulting in a finished protein

codon: triplet of bases in mRNA that codes for a specific amino acid which is carried by tRNA

coenzyme: a small, non-protein molecule which combines temporarily with an enzyme, allowing a reaction to proceed

competitive inhibitor: substance which binds reversibly to the active site of an enzyme, thus reducing the quantity of active enzyme available

complementary: referring to bases, the pattern of pairing adenine with thymine/uracil and cytosine with guanine

condom: artificial barrier method of contraception consisting of a thin rubber sheath placed over the erect penis preventing sperm from reaching ova

connective tissue: creates a framework for support and structure in organs and tissues

contextual cue: a way of helping retrieve a memory when the context of the encoding and the retrieval are the same

continuous fertility: ability of a male to produce viable sperm at a constant level during his reproductive life

contraception: any method used to prevent fertilisation of an ovum by a sperm or the successful implantation of a zygote or birth itself

convergent neural pathway: coming together of two or more neurons acting on one single neuron

corpus callosum: band of nerve tissue connecting the right and left hemispheres of the cerebrum

corpus luteum: structure formed from follicle capable of secreting progesterone

cyclical fertility: ability of a female to produce viable ova within a limited few days each month

cytokines: substances which are secreted by certain cells of the immune system and which have an effect on other cells

Glossary

dating scan: an ultrasound examination which is performed in order to establish the stage of a pregnancy and thereby estimate the delivery date of the baby

deep vein thrombosis: formation of a clot in vein in the leg

dehydrogenase enzyme: enzyme which catalyses the removal of hydrogen from a substrate

deletion: a base deleted from the sequence

dendrite: fibre in a neuron which carries nerve impulses towards cell body

desensitisation: a decreased response to a drug due to the formation of fewer and less sensitive neurotransmitter receptors

diaphragm: artificial barrier method of contraception which is inserted into the vagina to prevent sperm reaching ova

diastole: phase of the cardiac cycle when the heart muscle relaxes

differentiate: when a cell becomes a specific, more specialised type of cell through gene expression

divergent neural pathway: one single neuron generates an impulse which causes two or more subsequent neurons to fire

DNA polymerase: enzyme which adds complementary nucleotides to the deoxyribose (3′) end of a primer and newly forming DNA strand

dopamine: inhibitory neurotransmitter produced in the brain which blocks synaptic transmission and is strongly linked to reward mechanisms in the brain

double-blind trial: a clinical trial in which neither the patients nor the researchers know which subject is getting the vaccine or drug or which subject is getting the placebo

drug testing: testing potential medications to treat disease

duplication: genes are copied and remain in the chromosome

elaboration: any strategy for adding detail to a memory item to make it more likely to be stored

elastic fibres: fibres present in the wall of arteries, arterioles, veins and venules that allow the wall to stretch and recoil to accommodate the surge of blood after each contraction of the heart

electrocardiogram: trace of the electrical activity of the atria and ventricles of the heart as they contract and relax in one heartbeat

electron transport chain: series of carrier proteins attached to the inner membrane of the mitochondrion

embolus: mobile blood clot

embryo: an unborn or unhatched offspring in the process of development

embryonic stem cell: stem cell from the early embryo that self-renews and grows under the right conditions in a lab

encoding: converting incoming information into a form which the brain can then store to be available later

endometrium: lining of the uterus

endorphin: morphine-like chemical produced by the brain which plays an important role in pain reduction

endothelium: layer of cells which lines the inside of a blood vessel

energy investment phase: stage where ATP is invested into glycolysis

energy pay-off stage: stage where ATP is produced during glycolysis

epithelial cells: cell which lines the cavities and surfaces of blood vessels and organs throughout the body

ethical issue: issue arising from a set of principles regarding moral values and appropriate conduct

excitatory synapse: junction which promotes the continuance of a nerve impulse

exon: coding sequence of DNA

expressing: the process by which information from a gene is used in the synthesis of a functional gene product

fast twitch: muscle fibres which contract quickly but fatigue easily

fatigue: a feeling of tiredness due to accumulation of lactate

feedback inhibition: a mechanism used by a cell in which an enzyme that catalyses the conversion of a substrate into a product becomes inhibited when that product accumulates

fermentation: breakdown of glucose in the absence of oxygen

fertile period: narrow window of around 3–5 days when a woman can conceive

fertility: ability of sperm to fertilise an ovum and create a viable zygote

fibrin: tough protein formed by the action of thrombin on fibrinogen

fibrinogen: soluble plasma protein made in the liver

follicle: small cellular sac which has secretory abilities

follicle-stimulating hormone (FSH): hormone secreted by the pituitary gland, stimulating the growth of the follicles in the ovary in females. Promotes sperm production in males

follicular phase: stage in the menstrual cycle in which the follicles develop and their secretions start

frame-shift mutation: insertion or deletion of nucleotides which results in every subsequent codon to the right of the mutation in the base sequence being different and has a major effect on the structure of the protein produced

gamete: a sex cell, either sperm or ovum, which possesses only half the diploid number of chromosomes

gene expression: the translation of a base sequence in a gene into an amino acid sequence, and finally a finished protein, through the processes of transcription and translation

gene: the basic physical and functional unit of heredity

genetic code: base sequence of DNA

genome: entire hereditary information encoded in the DNA

genomic sequencing: ordering the sequence of nucleotide bases in a genome

genotype: the alleles or 'forms of a gene' an organism possesses

germline cell: a cell which can give rise to a gamete

glial cells: cells found in the nervous system which support neurons as well as producing myelin sheaths

glycolysis: initial series of reactions of cell respiration which takes place in the cytoplasm, with or without the presence of oxygen

haploid gamete: a gamete possessing a single set of unpaired chromosomes

herd immunity threshold: the number of people who need to be immunised in order to achieve herd immunity

high-density lipoprotein (HDL): lipoprotein which acts as a carrier for cholesterol

histamine: secreted by mast cells – causes vasodilation and increased capillary permeability

hydrogen bond: weak bond between the base pairs which holds both sides of a DNA molecule together

homologous chromosome: chromosome that has a partner which is the same size and shape and has the same sequence of genes present

human immunodeficiency virus: virus which causes acquired immune deficiency syndrome

hypertension: blood pressure above normal for a particular age group and health status

hypothalamus: region of brain associated with regulating body temperature

immunodeficiency: a state in which the immune system's ability to fight infectious disease and cancer is compromised or entirely absent.

in vitro: an artificial environment outside a living organism. For example, a laboratory

in vitro fertilisation (IVF): causing fertilisation event to occur outside the body of the female in a laboratory

induced fit: the ability of an enzyme to alter slightly the 3D shape of the active site in order to better fit the substrate after the substrate binds

inflammatory response: the complex biological response of body tissues to harmful stimuli such as pathogens, damaged cells or irritants

influenza: virus which causes respiratory infection commonly known as 'flu'

inhibitory synapse: junction which inhibits the continuance of a nerve impulse

insertion: an additional base inserted into the sequence

interstitial cell: cell found between seminiferous tubules which produce the male hormone testosterone

interstitial cell-stimulating hormone (ICSH): in males promotes the secretion of testosterone by the testes

intracytoplasmic sperm injection (ICSI): injection of a single sperm head into an ovum

intra-uterine device (IUD): artificial barrier method of contraception which consists of a small plastic or copper structure inserted into the uterus

intron: non-coding sequence of DNA

inversion: mutation in which a section of a chromosome is reversed

Glossary

karyotype: a way of representing the chromosomes found in a somatic cell nucleus classified according to their size and shape

lactate: compound formed as an end-product of anaerobic respiration during strenuous exercise

ligase: enzyme used to join fragments of DNA together during DNA replication and enzyme used to seal the gene into a plasmid in recombinant DNA technology

lipoprotein: water-soluble protein combined with fat found in the plasma

long-term memory: storage function in the brain for memories which have been processed and deemed important to keep

low-density lipoprotein (LDL): lipoprotein which acts as a carrier for cholesterol

luteal phase: stage in the menstrual cycle when the corpus luteum forms in the ovary and secretes progesterone, promoting endometrial development

luteinising hormone: in females triggers ovulation and the development of the corpus luteum

lymph: a clear to white fluid consisting of white blood cells, mainly lymphocytes

lymphocyte: type of white blood cell. They attack bacteria in the blood

mast cell: cell which releases histamine during inflammatory and allergic responses

matrix of the mitochondria: the fluid filled interior of the mitochondria

mature mRNA transcript: strand of messenger RNA that has had all introns removed and exons spliced together

meiosis: cell division which produces gametes

memory cell: a long-lived lymphocyte capable of responding to a particular antigen on its reintroduction, long after the exposure that prompted its production

menstrual cycle: events surrounding release of an ovum followed by changes in the lining of the uterus, usually taking 28 days to complete

menstruation: loss of the lining of the uterus along with some blood

metabolism: all the chemical reactions that occur within a living organism

missense mutation: a base is substituted for another resulting in a single amino acid change in the sequence or no change in the amino acid sequence

mitosis: a process where a single cell divides into two identical daughter cells

model cell: cell which has similar properties to a normal cell, within which the progress of a disease can be monitored

morning after pill: artificial chemical method of contraception using high levels of synthetic versions of progesterone and oestrogen

motor cortex: the part of the cerebral cortex in the brain in which originates the nerve impulses that initiate voluntary muscular activity

multipotent cell: cell that can develop into more than one cell type, but is more limited than a pluripotent cell

mutation: a change in the genetic composition of a cell

myelin: white phospholipid material which is the main constituent of the sheath around many neurons

myocardial infarction: heart attack caused by an embolism blocking up the coronary circulation

myoglobin: an iron-binding and oxygen-binding protein found in the muscle tissue of vertebrates and almost all mammals

NAD: carrier molecule which accepts hydrogen ions and electrons

negative feedback control: type of corrective mechanism for restoring and maintaining the dynamic state of an organism's internal environment in which a departure from a set value for a variable is detected and a response made to reduce the intensity of the increasing stimulus

net gain: output minus the input

neurotransmitter: chemical which acts as a communication between one nerve fibre and another across the synaptic cleft

non-coding: sequence of DNA bases that do not enter transcription and translation. No protein is synthesised

non-competitive inhibitor: substance which binds irreversibly to an area other than the active site of an enzyme, causing a change in the shape of the active site so that the normal substrate no longer fits

nonsense mutation: a substitution of one base for another resulting in a premature stop codon being produced

nucleotide: building block of the backbone of DNA. Composed of a base, a five-carbon sugar (ribose or deoxyribose) and at least one phosphate group

oestrogen: hormone produced in ovaries which promotes the development of female secondary sexual characteristics and during the menstrual cycle helps develop a suitable environment for an embryo to develop

oral contraceptive pill: artificial chemical method of contraception using synthetic versions of the naturally occurring hormones progesterone and oestrogen, which prevent the secretion of FSH and LH by the pituitary

ovulation: the release of an ovum from a follicle in the ovary. It usually occurs around the mid-point of the menstrual cycle

oxaloacetate: intermediate compound that joins with acetyl to form citrate

oxygen debt: a temporary oxygen shortage in the body tissues arising from exercise

parasympathetic nervous system: branch of autonomic nervous system which slows down processes in the body

pathogen: a bacterium, virus or other microorganism that can cause disease

peptide bond: type of bond formed between amino acids

peripheral nervous system (PNS): all of the nervous system with the exception of the central nervous system

peripheral vascular disease: damaged blood vessels distant from the heart

permeability: ability of cells to allow substances to pass in and out

personalised medicine: treatment which is based upon an individual's own genome

phagocyte: a type of cell capable of engulfing and absorbing bacteria and other small cells and particles

phagocytosis: the ingestion of bacteria by phagocytes

pharmacogenetics: the study of inherited genetic differences in drug metabolic pathways which can affect individual responses to drugs, both in terms of therapeutic and adverse effects

phenotype: physical characteristics of an organism, determined by the genotype

phenylketonuria: inherited error of metabolism due to a lack of an enzyme which can cause brain damage in a developing baby

phosphorylation: addition of a phosphate group to a molecule

pituitary gland: hormone-secreting gland, located in the base of the brain, which produces hormones that control many functions of other endocrine glands

placebo: a prescribed medicine that has no physiological effect

placebo-controlled: protocol used in drug testing using an agent similar to the drug but which has no physiological effect

plasma: straw-coloured liquid part of blood in which cells are suspended

pluripotent: term used to describe cells in the early embryo that can be grown in the laboratory and have the potential to differentiate into any type of cell

polymerase chain reaction (PCR): method by which multiple copies of a DNA fragment can be made in the laboratory; PCR is used to amplify a DNA fragment so that enough genetic material can be analysed to help solve crimes, help settle paternity suits and help diagnose genetic disorders

polypeptide: chain of many amino acids linked together by peptide bonds

postnatal screening: use of diagnostic checks after a baby is born to look for any abnormalities

pre-implantation genetic diagnosis (PGD): procedure used prior to implantation to help identify possible genetic defects within embryos often created using in vitro fertilisation

pressure filtration: pressure on capillary walls forces the filtration of plasma through the capillary wall, out of the circulatory system and into the intracellular space. The liquid surrounding the cells is now known as tissue fluid

primer: a short sequence of nucleotides that are complementary to specific target sequences at the two ends of a region of DNA to be amplified

progesterone: hormone produced in the ovaries which promotes development of the uterus wall and implantation as well as inhibiting further development of the follicle

Glossary

progesterone-only (mini) pill: artificial chemical method of contraception which uses only the synthetic version of progesterone

prostate gland: gland which surrounds the neck of the bladder in males and which secretes fluid components of semen

prothrombin: plasma protein made in the liver

puberty: time when a young person reaches sexual maturity and is capable of reproduction

pyruvate: formed from the splitting of a glucose molecule during the first stage of cell respiration called glycolysis; this occurs in the cytoplasm of the cell, with or without the presence of oxygen

recreational drug: chemical taken for pleasure or enjoyment but not under medical supervision

regulation: a rule made and maintained by an authority

rehearsal: a way of retaining information in the short-term memory by repetition

releaser hormone: hormone produced by the hypothalamus, which stimulates production of follicle-stimulating, luteinising and interstitial cell stimulating hormones

replicate: in the context of DNA, to make an exact copy

retrieval: process of recalling information from long-term memory

reverberating pathway: chain of neurons along which an impulse travels and is maintained by impulses generated by a neuron ahead

ribosome: structure found in the cytoplasm where a mature mRNA transcript is translated into a polypeptide chain

RNA polymerase: the enzyme that controls the synthesis of the mRNA strand during the transcription process in protein synthesis; it adds RNA nucleotides onto the 3' end of the newly forming mRNA strand

RNA splicing: process by which mature RNA transcript is formed; introns are removed and exons are spliced together

secondary exposure: second time that the pathogen infects

secondary tumour: occurs when cancer cells break away from the primary tumour and travel through the blood system to another part of the body

secretion: process by which substances are produced and discharged from a cell, gland or organ for a particular function in the organism, or for excretion

semen: collective name for sperm and the associated fluids produced by accessory glands

semilunar valve: structure preventing the backflow of blood from the aorta or pulmonary artery into the ventricles

seminal vesicle: one of a pair of glands on either side of the bladder in males which secrete fluid to nourish sperm and promote contraction of the female reproductive tract

seminiferous tubule: threadlike tubule within the testes which produces sperm from the epithelial cells which line the tubules

sensitisation: result of prolonged exposure to an antagonistic drug causing the individual to crave more of the drug

sensory cortex: the part of the cerebral cortex in the brain which receives incoming nerve impulses from sense organs

sensory memory: any memory that preserves the characteristics of a particular sensation

serial position effect: in a list of items presented to the short-term memory, those presented first and last are more likely to be recalled correctly than those in the middle

short-term memory: storage function in the brain for memories which have not received much processing

sino-atrial node: pacemaker of heart consisting of a cluster of cells located in the wall of the right atrium which controls the rate of heartbeat

slow twitch: muscle fibres which contract slowly but are less subject to fatigue

somatic cell: any cell within an animal which is not a gamete

somatic nervous system (SNS): nervous system associated with voluntary and reflex actions

specialised: differentiated cells with a specific function and which are unable to become any other kind of cell

specific immune response: specialised immunity for particular pathogens

sphygmomanometer: instrument for measuring blood pressure

splice-site: the boundary between an exon and an intron

start codon: first triplet which codes for the start of a polypeptide chain formation

statin: drug prescribed to help reduce high blood cholesterol levels

stem cell: an undifferentiated cell with no specific function, which has the potential to differentiate into different types of specialised cells

sterile: incapable of reproduction

sterilisation: artificial barrier method of contraception which involves surgery to close and seal the sperm ducts preventing the release of sperm

stop codon: final triplet which codes for termination in polypeptide chain formation

storage: retention of information so that it can be recalled later

stroke volume: volume of blood pumped by one of the ventricles during a single contraction

substitution mutation: type of gene mutation in which one nucleotide is substituted for another in a DNA sequence; there are three types of substitution mutation: missense, nonsense and splice-site

substrate: molecule upon which an enzyme acts

summation: additive effect of two or more weak stimuli to cause a nerve impulse to occur

super ovulation: release of more than one ovum as a result of hormone treatment

sympathetic nervous system: branch of autonomic nervous system which speeds up processes in the body

synapse: a junction between two nerve cells, consisting of a minute gap across which impulses pass by diffusion of a neurotransmitter

synaptic cleft: microscopic gap between neurons

systole: phase of the cardiac cycle when the heart muscle contracts

testosterone: male sex hormone manufactured by the interstitial cells which stimulates sperm production as well as the development of male sexual characteristics and prostate and seminal vesicle function

therapeutics: the branch of medicine concerned with the treatment of disease

thermal cycler: automated machine able to carry out repeated cycles of PCR by varying temperature

thrombin: enzyme which converts fibrinogen into fibrin to form a clot

thrombus: blood clot

tissue fluid: liquid formed when plasma is filtered through the capillary walls containing no cellular components or proteins

T lymphocyte: white blood cell which destroys infected body cells by recognising antigens on the cell membrane and inducing apoptosis

tolerance: when the effect of a drug becomes progressively diminished unless the dosage is increased

transcription: first stage in protein synthesis (gene expression) that occurs in the nucleus of the cell and results in the production of a strand of mRNA called the 'primary transcript'

translation: the second stage in gene expression where the mature mRNA transcript is translated into an amino acid sequence using tRNA molecules

translocation: a chromosome mutation resulting from the addition of a section of a chromosome to another chromosome that is not its homologous partner

tumour: a swelling caused by an abnormal growth of tissue, either malignant or benign

ultrasound imaging: use of high frequency sounds to create images of internal structures

unspecialised somatic cells: undifferentiated cells capable of becoming any type of cell

vaccination: the administration of antigenic material (a vaccine) to stimulate an individual's immune system in order to develop adaptive immunity to a pathogen

vasoconstriction: narrowing of the arteries as a result of the contraction of the smooth muscles found in the walls, increasing blood flow

vasodilation: widening of the arteries as a result of the contraction of the smooth muscles found in the walls, reducing blood flow

venule: a small division of a verin

vesicle: general term for any sac-like structure which contains fluid

Answers to Area Revision Tests

Area 1 Revision Test

1. (a) Stem cell – unspecialised somatic cell that can divide to make copies of itself (self-renew) and/or differentiate into specialised cells (1)

   Tumour – a mass of abnormal cells (1)

   Germline cell – gamete, such as sperm or ovum (1)

   (b) The repair of damaged or diseased organs or tissues or a specific example of this (1)

2. (a) Hydrogen bond (1)

   (b) Y = phosphate,   Z = deoxyribose sugar (2)

   (c) Guanine (1)

   (d) Adenine and thymine (2)

   (e) Antiparallel (1)

3. True (1)

   False – thymine (1)

   False – ribose (1)

4. (a) X1 = exon, Y1 = intron (2)

   (b) RNA splicing (1)

   (c) Nucleus (1)

   (d) tRNA (1)

5. Substitution – no (1)

   Insertion – yes (1)

   Deletion – yes (1)

6. (a) Amplification of DNA (1)

   (b) Separate strands (1)

   (c) Short sequences of DNA which start DNA replication (1)

   (d) Heat-tolerant DNA polymerase (1)

   (e) 1024 (1)

7. (a) Anabolic (1)

   (b) Competitive inhibition (1)

   (c) Decrease (1)

   (d) Feedback inhibition (1)

8. (a) Citric acid cycle (1)

   (b) Pyruvate (1)

   (c) Inhibitory effect / reduces / decreases rate (1)

9.

Feature	Type of skeletal muscle fibre	
	Slow twitch (type 1)	Fast twitch (type 2)
Activity most suitable for	Endurance activities	Sprinting, short bursts
Myoglobin present	Yes	No
Major storage fuels	Fats	Glycogen
Contraction speed	Slower	Faster
Contraction duration	Longer	Shorter

(1 for each correct line in table)

10. (a) As the concentration of glucose increases the rate of respiration also increases (1)

(b) 67% (1)

(c) 22 seconds (1)

(d) Between 2% and 3% (1)

Area 2 Revision Test

1. (a) Corpus luteum (1)

    (b) 1 (1)

    [The most mature follicle will be the one just before ovulation takes place.]

    (c) They produce the hormones oestrogen (1) and progesterone (1)

2. (a) Releaser (1)

    (b) Interstitial cell-stimulating hormone (ICSH) (1)

    (c) (i) Follicle-stimulating hormone (FSH) (1)

    (ii) Promotion of sperm production (1)

    (d) As the levels of testosterone rise the pituitary secretes less ICSH (1) and FSH (1)

3. (a)

True	False	Correction
✔		
	✔	chemical
✔		

(1 for each correct line in table)

    (b) (i) A (1)

    [Calculate the number of sperm/cm^3 for each man by dividing the number of sperm by the volume of semen.]

    (ii) C (1)

    [This male's sperm count/cm^3 is 17·2, which is below the World Health Organization's definition of normal.]

    (c) 600–1200 ng/100 cm^3 (1)

    [A range means the lowest to the highest.]

4. (a) Age of fetus **or** risk of genetic disorder (1)

    (b) An anomaly scan is used to detect serious physical abnormalities in the fetus (1)

    (c) It can be carried out relatively early in pregnancy (1)

    [Note the advantages and disadvantages of each of the screening procedures which are available to a pregnant woman.]

5. (a) X^AX^a (1)

    [Since she is unaffected, she must have at least one dominant allele. But she has a son who is affected and he must have picked up a defective allele on the X chromosome he inherited from his mother, who must therefore be heterozygous or a carrier.]

    (b) Mother must be heterozygous X^AX^a and father is unaffected X^AY (1)

    (c) 1:1 (1)

    (d) The sample size is too small **or** the random nature of fertilisation (1)

    [Frequently the expected outcome and the actual outcome are different due to these variables.]

6. (a) Elastic fibres (1)

    (b) Tissue fluid (1)

    [This fluid is forced out of the capillaries by pressure filtration and bathes tissue cells.]

(c) Thin walled (1)

[Remember this applies in other contexts, such as the wall of the alveolus in the lung.]

(d) The central lumen of an artery is much narrower than that of a vein (1)

(e) Tissue fluid does not contain proteins whereas blood plasma does (1)

7. (a) (i) X (1)

(ii) Sino-atrial node (SAN)/pacemaker (1)

(iii) Auto-rhythmic (1)

(b) Atrial systole (1)

[As the left atrium contracts, the pressure forces the atrio-ventricular valve open.]

8. (a) Atheroma (1)

(b) Atheroma grows. Diameter of lumen becomes reduced and blood flow is restricted resulting in increased blood pressure (2)

[Anything which slows down the flow of blood in a vessel will increase the pressure.]

9. (a) (i) Pancreas (1)

(ii) As glucose levels rise above normal, pancreatic receptors respond by increasing secretion of insulin; this activates the conversion of glucose to glycogen in the liver thus decreasing blood glucose concentration (2)

or

As glucose levels fall below normal, pancreatic receptors respond by increasing secretion of glucagon by the pancreas. Glucagon activates the conversion of glycogen to glucose in the liver thus increasing blood glucose concentration (2)

(b) A BMI of greater than 30 indicates obesity/high body fat content in relation to both height and weight/calculated by dividing the body mass [kg] by the square of the height [m^2] (any two) (2)

(c) (i) 2:5 (1)

[Express the ratio of 20:50 as a simple whole number ratio.]

(ii) 80% (1)

[Prediction assumes the same rate of increase as for previous years of 15%.]

Area 3 Revision Test

1. (a) Cerebral cortex (1)

    The corpus callosum connects both halves of the cerebrum/cerebral hemispheres (1)

    [The terms cerebrum and cerebral hemispheres are often used interchangeably.]

    Medulla (1)

    [Many autonomic centres are located here.]

    (b) (i) Midbrain (not mesencephalon) (1)

    (ii) Telencephalon **and** forebrain (2)

2. (a) Generates a large surface area (1)

    (b) Corpus callosum (1)

3. (a) Any two from:

    Rehearsal: repetition of information over and over again

    Organisation: adding structure to information/categorising information

    Elaboration: adding meaning to information/analysing the meaning of information (2)

    (b) Involves grouping smaller pieces of information into single items / increases span of short-term memory (1)

    (c) Stimulus which helps retrieve information from long-term memory / a link to the time and place in which a memory was created (1)

4. (a) Axon (1)

    (b) Speeds up the rate of transmission of impulse / insulates the nerve fibre (1)

    (c) (i) Vesicles (1)

    (ii) Pain perception would be diminished (1)

    [If the transmission of an impulse is blocked, the next neuron(s) will not fire so the perception of the pain is reduced.]

5. (a) (i) Epithelial cells (1)

    (ii) Histamine (1)

    (iii) Vasodilation / increased capillary permeability (1)

    (b) Phagocytes recognise pathogens. They engulf them and use digestive enzymes to destroy them (1)

6. (a) T lymphocyte – induces apoptosis (1)

    B lymphocyte – produces antibodies (1)

    Antigen – surface protein which triggers a specific immune response (1)

    (b) Failure of the regulation of the immune system leads to T lymphocytes responding to self-antigens. (1)

7. true

    false B lymphocyte

    true

8. (a) 2012 (1)

   (b) 100% (1)

   (c) As the number of people living with HIV increases, the number of people accessing treatment increases/the number of people living with AIDS is still increasing each year (2)

   (d) HIV, the major cause of AIDS, attacks and destroys T lymphocytes (1)

9. (a) Herd immunity occurs when a large percentage of a population is immunised. Non-immunised individuals are protected as there is a lower probability they will come into contact with infected individuals (1)

   (b) Malnutrition, poverty or vaccine rejected by a percentage of the population (1)

   (c) The mechanism by which a pathogen alters its surface proteins in order to evade a host immune response (1)

10. (a) Clinical trials are randomised in order to reduce bias (1)

   (b) One group of subjects receive the vaccine or drug while the second receive a placebo control but neither the subjects nor the researchers are aware of which group receives the drug (1)

   (c) Low numbers of subjects can result in a high level of experimental error when interpreting results (1)

Index

A *see* adenine

acetyl coenzyme A 33, 116

acetylcholine 68, 96, 116

acquired immune deficiency syndrome (AIDS) 107, 116

activation energy 29, 116

active sites 29, 116

adenine (A) 11, 12, 17

adenosine diphosphate (ADP) 32–34

adenosine triphosphate (ATP) 13, 32–6, 41

  see also ATP synthase

adjuvants 108, 116

adrenal gland 74, 116

adrenaline 74

aerobic respiration 36

affinity 29, 116

agonistic drugs 99, 101, 116

AIDS *see* acquired immune deficiency syndrome

albinism 56

alleles

  autosomal defective 57

  autosomal defective dominant 57

  defective sex-linked recessive 57

allergies 106, 116

amino acids 18–24

  and mutation 23–4

  sequence 21

amniocentesis 54–5, 116

amniotic fluid 54, 116

amplification 15, 116

anabolic 28, 116

anaemia

  sickle-cell 57

anomaly scan 53, 116

ANS *see* autonomic nervous system

antagonism 68, 87, 116

antagonistic drugs 99, 101, 116

antagonists 99, 101

antenatal (prenatal) screening 53, 82–3, 116

  biochemical testing 54

  diagnostic testing 54–5

  ultrasound imaging 53

antibodies 105–6, 116

anticodons 18, 116

antigen–antibody complexes 106, 116

antigenic variation 110, 116

antigens 104–8, 110, 116

antiparallel 12, 14, 116

apoptosis 105, 116

arteriole 61, 116

artery 59–60, 83–4

  blockage 70

  narrowing 72

arthritis, rheumatoid 105

artificial insemination 50, 116

association areas 90–1, 116

atheroma 70–4, 116

atherosclerosis 70, 72, 74

ATP (adenosine triphosphate) 13, 32–6, 41, 116

  net gain 32

ATP synthase 34, 116

atrio-ventricular node 66, 116

atrio-ventricular valve 64, 116

atrium 59, 63, 66

autoimmune diseases 105

autoimmune system 105

autonomic nervous system (ANS) 68, 87, 116

autosomal defective alleles 57

autosomal dominant defective inheritance 57, 73

autosomal recessive defective inheritance 56

autosome 116

axons 95, 117

B lymphocytes 105–6

barriers 52, 117

bases

Index

Index

Index

Index

Leckie
the education publisher
for Scotland

Higher
HUMAN
BIOLOGY

For SQA 2019 and beyond

Practice Papers

John Di Mambro, Deirdre
McCarthy and Stuart White

Revision advice

The need to work out a plan for regular and methodical revision is obvious. If you leave things to the last minute, it may result in panic and stress which will inhibit you from performing to your maximum ability. If you need help, it is best to find this out when there is time to put it right. Revision planners are highly individual and you need to produce one that suits you. Use an area of your home that is set aside only for studying if possible, so that you form a positive link and in this way will be less liable to distractions. Not only do you need a plan for revising Higher Human Biology, but also for all your subjects. Below is one revision plan but you will have your own ideas here!

Work out a revision timetable for each week's work in advance – remember to cover all of your subjects and to leave time for homework and breaks. For example:

Day	6.00–6.45 pm	7.00–8.00 pm	8.15–9.00 pm	9.15–10.00 pm
Monday	Homework	Homework	English revision	Human Biology revision
Tuesday	Maths revision	Physics revision	Homework	Free
Wednesday	Geography revision	English revision	Human Biology revision	Maths revision
Thursday	Homework	Physics revision	Geography revision	Free
Friday	English revision	Human Biology revision	Free	Free
Saturday	Free	Free	Free	Free
Sunday	Maths revision	Physics revision	Geography revision	Homework

Make sure that you have at least one evening free each week to relax, socialise and re-charge your batteries. It also gives your brain a chance to process the information that you have been feeding it all week.

Arrange your study time into sessions that suit you, with a 15-minute break in between. Try to start studying as early as possible in the evening, when your brain is still alert, and be aware that the longer you put off starting, the harder it will be.

If you miss a session, do not panic. Log this and make it up as soon as possible. Do not get behind in your schedule – discipline is everything in being a successful student.

Study a different subject in each session, except for the day before an exam.

Do something different during your breaks between study sessions – have a cup of tea, or listen to some music. Do not let your 15 minutes expand into 20 or 25 minutes!

Have your class notes and any textbooks available for your revision to hand, as well as plenty of blank paper, a pen, etc. If relevant, you may wish to have access to the Internet but be careful you restrict using this only for supporting your revision. You may also like to make keyword sheets like the example below:

Keyword	Meaning
Ribosome	Structure in the cell that manufactures protein
Pathogen	Agent, such as a virus or bacterium, that is capable of causing disease

Flashcards are another excellent way of practising terms and definitions. You can make these easily or buy them very cheaply. Use flashcards either to recall the keyword when you see the meaning, or to give the meaning when you see the keyword. There are several websites that are free to use and give you the ability to generate flashcards online. If you collaborate with your friends and take different sections of the course, you can merge these into a very powerful learning and revision aid.

Finally, forget or ignore all or some of the advice in this section if you are happy with your present way of studying. Everyone revises differently, so find a way that works for you!

Command words

In the papers and in the Higher exam itself, a number of command words will be used in the questions. These command words are used to show you how you should answer a question: some words indicate that you should write more than others. If you familiarise yourself with these command words, it will help you to structure your answers more effectively.

Command word	Meaning/explanation
Name, state, identify, list	Giving a list is acceptable here – as a general rule you will get 1 mark for each point you give.
Suggest	Give more than a list – perhaps a proposal or an idea.
Outline	Give a brief description or overview of what you are talking about.
Describe	Give more detail than you would in an outline, and use examples where you can.
Explain	Discuss why an action has been taken or an outcome reached – what are the reasons and/or processes behind it?
Justify	Give reasons for your answer, stating why you have taken an action or reached a particular conclusion.
Define	Give the meaning of the term.
Compare/contrast	Give the key features of **two** different items or ideas and discuss their similarities/differences.
Predict	Work out what will happen.

In the exam

Watch your time and pace yourself carefully. Work out roughly how much time you can spend on each answer and try to stick to this.

Be clear before the examination what the instructions are likely to be, for example how many questions you should answer in each section. The papers will help you to become familiar with the examination instructions.

Read the question thoroughly before you begin to answer it – make sure you know exactly what the question is asking you to do. If the question is in sections, make sure that you can answer each section before you start writing.

Plan your extended responses by jotting down keywords, making a brief mind-map of the important points or whatever you find works best for you.

Do not repeat yourself as you will not get any more marks for saying the same thing twice. This also applies to annotated diagrams, which will not get you any extra marks if the information is repeated in the written part of your answer.

Give proper explanations. A common error is to give descriptions rather than explanations. If you are asked to explain something, you should be giving reasons. Check your answer to an **explain** question, and make sure that you have used linking words and phrases such as **because**, **this means that**, **therefore**, **so**, **so that**, **due to**, **since** and **the reason is**.

Good luck!

Paper brief and question allocation by type

The paper brief on page 145 shows the breakdown of questions within the paper.

This will enable you to allocate your revision proportionately to maximise your chance of success. For example, with knowledge-based questions making up approximately 70% of the marks, you should spend approximately 70% of your time revising your notes and practising this type of question.

Knowledge-based questions are split into two areas: **demonstrating** knowledge and **applying** knowledge.

- Demonstrating involves mostly recall, whereas applying involves you using that knowledge in unfamiliar situations that you won't have learned.
- The best way to improve the application of knowledge is to practise from papers and textbooks.
- In these questions you will also be provided with context and it is important to read all the information provided in the question to allow you to identify the knowledge you need to recall or apply.

The **skills section**, approximately 30% of the marks, involves data handling and experimental design questions. The skills revision should take up approximately 30% of your revision time.

- The data handling questions, approximately 5–9 marks, give you the data to process or make predictions or conclusions. The best way to improve your technique with these questions is practice. The past papers and textbooks will provide you with the opportunity to do this.

Be sure to read all of the data and information carefully and, as always, read the question more than once. You are often asked to use only part of the data to answer a particular question.

- The **experimental design** questions, approximately 5–9 marks, give you lots of information about an experiment and you will be asked questions about the design or results provided. Once again, the key to answering successfully is to read all of the data and information carefully and to do as many practice questions as you can. The questions asked are often similar, but it is the experiment which differs. You should be aware that techniques such as PCR and gel electrophoresis and the apparatus used can be assessed here as well.

The table below details where these areas are covered within these papers, so if you are struggling with one area, you can do multiple questions to practise.

Component	Marks		
	Knowledge-based	Skills	Total
Question Paper	85±5	35±5	120

Knowledge-based	Marks
Demonstrating knowledge of human biology by making statements, describing information, providing explanations and integrating knowledge.	min 30
Applying knowledge of human biology to new situations, interpreting information and solving problems.	min 30

Skills	Marks
Planning and designing experiments/investigations.	30–40
Selecting information from a variety of sources.	
Presenting information appropriately in a variety of forms.	
Processing information/data (using calculations and units, where appropriate).	
Making **predictions** and generalisations based on evidence/information.	
Drawing valid **conclusions** and giving explanations supported by evidence/justification.	
Evaluating experiments/investigations and suggesting improvements.	

Question type	Question number	
	Exam A (Paper 2)	Exam B (Paper 2)
An extended writing question worth 10 marks with a choice from 2 questions.	16 A 16 B	14 A 14 B
One large data handling question: 5–9 marks.	14	6
One large experimental design question: 5–9 marks.	6	9
'A' type marks: questions set at grade A level of difficulty – approximately 30%.	Throughout paper, usually with keywords like 'explain' or 'describe' and worth more than 1 mark.	

Links to the syllabus

	Key area	Practice Paper P1 – Paper 1 P2 – Paper 2			
		Exam A		Exam B	
		P1	P2	P1	P2
Area 1 : Human cells Demonstrating and applying knowledge	Division and differentiation in human cells	1	1		4(a)(b)(c)
	Structure and replication of DNA	6	2		1
	Gene expression	7 8 9	3 5 (d)	2 3 23	2(b) (c) (e)
	Mutations		10 (a) (b)		
	Human genomics		10 (d)		3
	Metabolic pathways	10		5 7	
	Cellular respiration	2	5	8 9	
	Energy systems in muscle cells	11			5
Area 2 : Physiology and health Demonstrating and applying knowledge	Gamete production and fertilisation			10	
	Hormonal control of reproduction	12	13 (a) 16A		
	The biology of controlling fertility		10 (e) 13 (b)	11	7(c) (d) (e)
	Antenatal and postnatal screening			12 19	
	Structure and function of arteries, capillaries and veins	14	7		
	Structure and function of the heart		15(c)	13 15	10(d) 14B
	Pathology of cardio-vascular disease	15 16	8	14	
	Blood glucose levels and obesity		12(c)		6(b)

Key Area	Practice Paper P1 – Paper 1 P2 – Paper 2			
	Exam A		Exam B	
	P1	P2	P1	P2
Area 3: Neurobiology and immunology — Demonstrating and applying knowledge				
Divisions of the nervous system and neural pathways				
The cerebral cortex	19			
Memory	20	9(a) (b) (c)		11(g) (h)
The cells of the nervous system and neurotransmitters at synapses	21		20 21	14A
Non-specific body defences			22	13a
Specific cellular defences against pathogens	22			8 (b) 13(b) (c) (d)
Immunisation	24	16B (i)		
Clinical trials of vaccines and drugs	23	16B (ii)	23	4 (d) 12
Skills of scientific enquiry				
Selecting information		4 (a) 11 (a) (iii) (iv)	16	9(b) 10(a) (b)
Presenting information		6(e)		9(g) 11(a)
Processing information	3 4 5 13 17	10 (c) 11 (b) 12(a) (b) 14(a) 15(a) (b)	1 4 6 17	2(a) (d) 6(c) (e) (f) 7(b) 9(a) 10(b) 11(d)
Planning investigations		6(a) 4 (b) (c) (d) (i)	18	9(d) (e)
Evaluating experimental procedures		6 (b) (c) (d) 9 14 (c) (e)		8(a) 9(f) 10(c) 11(e) (f)
Drawing conclusions	18 25	6(f) 11 (a) (i) 14 (b)	24 25	9(h)
Predicting and generalising		4(d)(ii) 11(a)(ii) 14(d)		6(a) (d) 7(a) 9(c) 11(b) (c)

Practice Paper A

Higher Human Biology

Practice Papers for SQA Exams

Practice Paper A

Human Biology
Paper 1 — Multiple choice

Duration — 40 minutes

Total marks — 25

Attempt ALL questions.
You may use a calculator.

Leckie
the education publisher
for Scotland

PAPER 1 – 25 marks

Attempt ALL questions.

1. Which of the following descriptions is correct?

   A   Stem cells are somatic cells which are relatively unspecialised

   B   Somatic cells undergo mitosis to produce haploid cells

   C   Somatic cells undergo meiosis to produce diploid cells

   D   Germline cells can only undergo meiotic cell division.

2. Which of the following would not be found in the matrix of a mitochondrion?

   A   Glucose

   B   Adenosine triphosphate

   C   Acetyl coenzyme A

   D   Carbon dioxide

3. One form of cancer treatment involves chemotherapy, which uses a drug to kill dividing cells.

   The graph shows the number of healthy cells and cancer cells in the blood of a patient receiving chemotherapy. The arrows show when the drug was given to the patient.

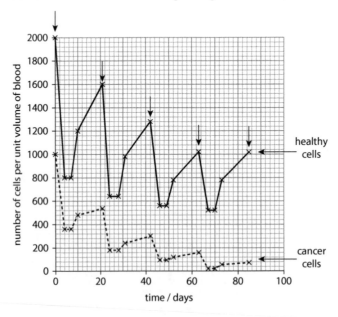

3. (continued)

Calculate the simplest whole number ratio of healthy cells to cancer cells, per unit volume of blood, at day 4.

A   2:1

B   20:9

C   80:36

D   400:180

4. If 20% of the bases in a molecule of DNA are thymine, what is the ratio of thymine to cytosine in the same molecule?

A   1:2

B   2:3

C   3:4

D   4:5

5. The concentration of urea in a healthy person's urine was found to be 60 times as much as found in the blood.

If a urine sample from such a person contained 6 g urea/litre, how much urea would be present in 100 cm³ of blood?

A   1·0 g

B   0·1 g

C   0·01 g

D   0·001 g

6. Which line in the table correctly describes the structure of DNA?

	Nucleotide contains	Backbone composed of
A	deoxyribose sugar, phosphate and base	hydrogen bonds
B	ribose sugar, phosphate and base	hydrogen bonds
C	deoxyribose sugar, phosphate and base	sugar–phosphate
D	ribose sugar, phosphate and base	sugar–phosphate

7. An RNA nucleotide could be formed from phosphate,

    A   ribose and uracil

    B   ribose and thymine

    C   deoxyribose and uracil

    D   deoxyribose and thymine.

8. Exons in the primary transcript of mRNA are

    A   non-coding regions and are joined together to form a mature transcript

    B   coding regions and are joined together to form a mature transcript

    C   non-coding regions and are removed in RNA splicing

    D   coding regions and are removed in RNA splicing.

9. Which of the following statements is correct with regard to eukaryotic RNA splicing?

    A   RNA splicing takes place in the cytoplasm

    B   the primary transcript contains only introns

    C   introns are retained, and exons are removed

    D   the nucleus is the site of RNA splicing.

10. Identify the correct description of enzyme inhibition.

    A   In competitive inhibition, the inhibitor binds to the active site of the enzyme

    B   Non-competitive inhibitors are very similar in structure to the enzyme's normal substrate

    C   In competitive inhibition, a poison binds to the enzyme so that it can never work again

    D   All inhibitors act in an irreversible fashion.

11. Which line in the table below correctly describes slow twitch muscle fibres?

	Contraction speed	Contraction duration
A	slowly	longer
B	quickly	longer
C	slowly	shorter
D	quickly	shorter

12. Which of the following statements correctly describes the role of follicle-stimulating hormone (FSH) in males?

A Stimulates the production of testosterone

B Promotes sperm production

C Activates the prostate gland and seminal vesicles

D Stimulates the pituitary gland.

13. The sperm counts of a sample of men taken between 1950 and 2010 are shown in the graph below.

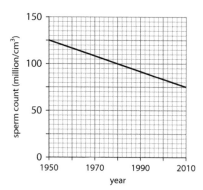

What is the average reduction in sperm count per year?

A 0·75 million/cm³/year

B 0·83 million/cm³/year

C 0·92 million/cm³/year

D 50 million/cm³/year

14. The diagram below shows a blood vessel.

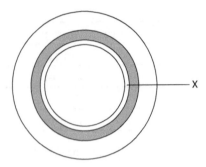

Which of the following is structure X?

A   Endothelium

B   Central lumen

C   Elastic fibre

D   Smooth muscle.

15. High blood cholesterol level is

A   lowered if the ratio of low-density to high-density proteins is high

B   linked to the development of atherosclerosis

C   unaffected by exercise

D   unrelated to diet.

16. Which of the following combinations would reduce the risk of developing cardiovascular disease (CVD)?

A   Keeping weight under control, minimising stress, reducing hypertension and improving HDL blood profiles

B   Keeping weight under control, increasing stress, increasing hypertension and improving HDL blood profiles

C   Losing lots of weight, minimising stress, reducing hypertension and increasing HDL blood profiles

D   Keeping weight under control, minimising stress, reducing hypertension and increasing HDL blood profiles.

17. The cell shown below is magnified 500 times. What is the actual width of the cell?

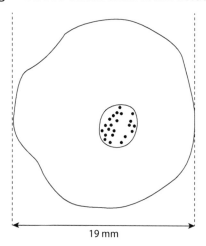

19 mm

A   1900 μm

B   95 μm

C   38 μm

D   3·8 μm

18. People with type 2 diabetes have cells with low sensitivity to insulin. About 80% of people with type 2 diabetes are overweight or obese. Some people who are obese have gastric bypass surgery (GBS) to help them to lose weight.

Doctors investigated whether GBS affected sensitivity to insulin. They measured patients' sensitivity to insulin before and after GBS. About half of the patients had type 2 diabetes. The other half did not but were considered at high risk of developing the condition.

The table shows the doctors' results. The higher the value in the table, the greater the sensitivity to insulin.

Patients	Mean sensitivity to insulin/arbitrary units	
	Before gastric bypass surgery	1 month after gastric bypass surgery
Did not have diabetes	0·55	1·30
Had type 2 diabetes	0·40	1·10

Which of the following conclusions can be drawn from the data **in the table**?

A   People with type 2 diabetes have a greater sensitivity to insulin after GBS than those who do not have type 2 diabetes

B   People with type 2 diabetes have a decreased sensitivity to insulin after GBS compared with those who do not have type 2 diabetes

C   People with type 2 diabetes are more likely to have GBS than those who do not have type 2 diabetes

D   People with type 2 diabetes are more obese than those who do not have type 2 diabetes.

19. Which of the following does **not** correctly describe the function of the cerebral cortex?

A   Receiving sensory information

B   Co-ordinating voluntary movement

C   Making decisions in the light of experience

D   Regulating the basic life processes of breathing, heart rate, arousal and sleep.

20. The diagram below shows the three levels of memory, X, Y and Z.

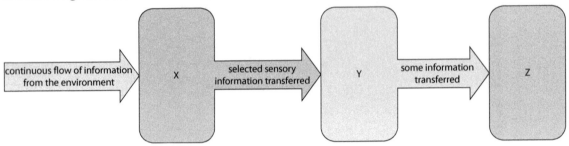

Select the correct type of memory for X, Y and Z.

	X	Y	Z
A	sensory	long-term	short-term
B	sensory	short-term	long-term
C	short-term	long-term	sensory
D	short-term	sensory	long-term

21. The diagram below represents a neural pathway. The arrows indicate the direction of nerve impulses.

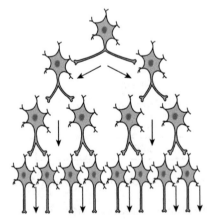

The type of pathway shown can be described as

A   sensory

B   converging

C   diverging

D   reverberating.

22. Which of the following correctly describes possible actions of T and B lymphocytes?

	T lymphocytes	*B lymphocytes*
A	destroy infected cells by inducing apoptosis	attack insulin-producing cells in type 1 diabetes (autoimmune response)
B	secrete antibodies into the lymph and blood where they make their way to the infected area	activated by antigen presenting cells
C	secrete cytokines	attack insulin-producing cells in type 1 diabetes (autoimmune response)
D	may become memory cells	secrete antibodies into the lymph and blood where they make their way to the infected area

23. Vaccine clinical trials

A   should be randomised to eliminate bias

B   need the researchers to know which group the subjects have been allocated to

C   do not use placebo controls

D   work best with a small number of subjects.

24. The influenza virus is difficult to eliminate because

A   it is unable to change its antigenic structure

B   memory cells are highly effective against it

C   the vaccine is not fully effective

D   the modified virus has lost its antigenicity.

25. Nutritionists investigated the relationship between eating oily and non-oily fish and the incidence of asthma. They analysed the diets of children with asthma and the diets of children without asthma.

The figure shows the results from their investigation.

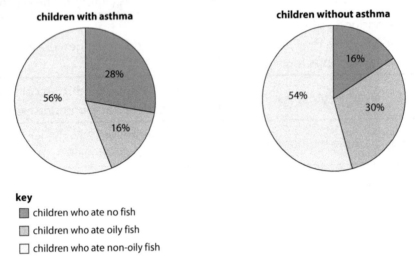

Which of the following conclusions is **not** correct, based only on the data **in the figure**?

A   It is less likely that children with asthma ate oily fish

B   It is less likely that children with asthma ate fish

C   There is only 2% difference in children with or without asthma who ate non-oily fish

D   28% of children without asthma ate oily fish.

Higher Human Biology

Practice Papers for SQA Exams

Practice Paper A

Human Biology
Paper 2

Duration — 2 hours 20 minutes

Total marks — 95

Attempt ALL questions.
You may use a calculator.
Question 16 contains a choice.

Leckie
the education publisher
for Scotland

PAPER 2 – 95 marks

Attempt ALL questions.

1. The diagram below shows a somatic cell undergoing mitosis.

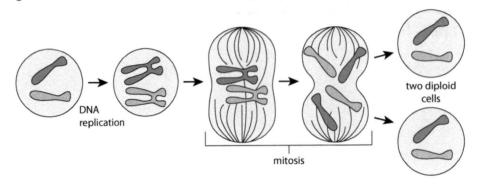

(a) State another type of cell which divides by mitosis but also by meiosis. 1

_____

(b) Stem cells are unspecialised somatic cells that can differentiate into specialised cells.

Name **two** types of stem cells. 1

_____

(c) Give **one** difference between these two types of stem cells. 1

_____

_____

(d) Describe what is meant by cellular differentiation. 1

_____

_____

(e) Give an account of the research and therapeutic uses of stem cells and the issues surrounding their use. 3

_____

_____

_____

_____

_____

1. (continued)

(f) Cancer cells divide excessively to produce a mass of abnormal cells called a tumour.

Describe what can happen if cancer cells fail to attach to each other. **1**

_____

_____

2. The following diagram shows the stages in the replication of DNA.

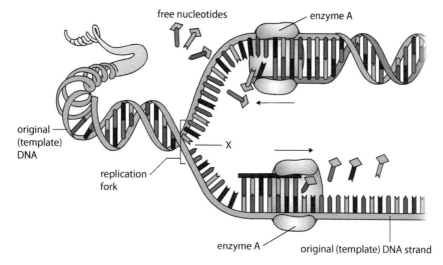

(a) Identify enzyme A. **1**

_____

(b) Using the numbers 1 and 2, identify **on the diagram** the strand of DNA which is replicated continuously and the strand of DNA which is replicated in fragments respectively. **2**

(c) Explain what is happening at point X. **1**

_____

_____

3. The diagram below shows the process of transcription during gene expression.

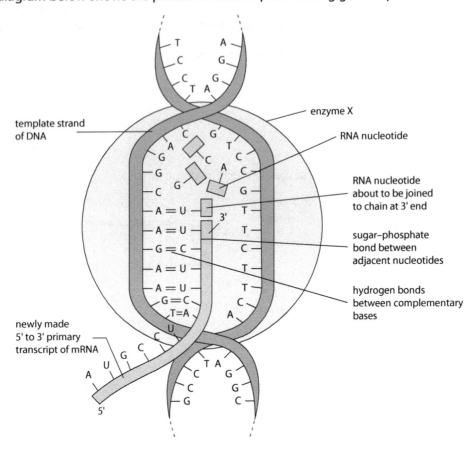

(a) Name the location within the cell where this process occurs. **1**

_____

(b) Name enzyme X. **1**

_____

(c) Describe the function of this enzyme. **2**

_____

_____

_____

_____

3. (continued)

(d) The diagram shows that RNA has the base uracil, while DNA has the base thymine.

State **one** other difference between RNA and DNA.   1

_____

_____

(e) The primary transcript of RNA can then undergo a process.

State the name of this process and give an account of it.

Process _____   1

Account   2

_____

_____

_____

4. Large molecules, such as nucleic acids, can be separated using the technique of gel electrophoresis.

(a) The following stages take place in this procedure for two samples of DNA:

1   apply current

2   inject DNA ladder into first well

3   inject samples into second and third wells

4   DNA moves towards positive anode

Arrange these stages, starting with the first, in the correct order.

_____ → _____ → _____ → _____   2

(b) Name the apparatus used to

(i) apply the current _____   1

(ii) inject the wells with the DNA _____   1

(c) Explain why the DNA is held in a buffer.   1

_____

_____

4. (continued)

(d) The diagram below shows the result of the gel electrophoresis.

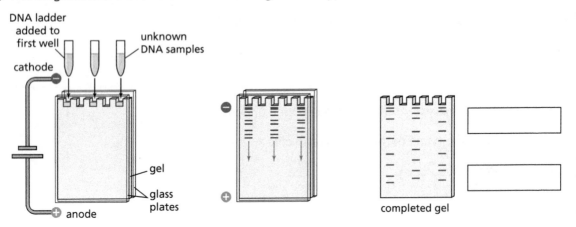

(i) Explain why a DNA ladder is inserted into the first well. **2**

_____

_____

(ii) In **one** of the blank boxes shown at stage 3, insert the words "large fragments". **1**

5. The diagram below shows the transfer of chemical energy between metabolic pathways.

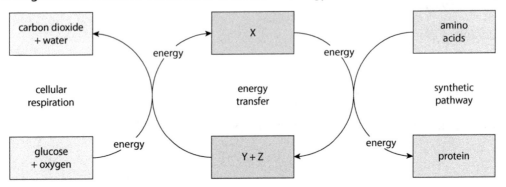

(a) Name substances X, Y and Z. **2**

X _____   Y _____

Z _____

(b) Name the type of metabolic pathway that releases energy, as represented by cellular respiration in this diagram. **1**

_____

(c) Describe the role of oxygen after ATP has been synthesised. **1**

_____

5. (continued)

(d) The amino acids in the synthetic pathway are joined to form a chain.

Name the bond which joins amino acids to one another.   **1**

_____

6. An experiment was carried out to investigate the effect of different respiratory substrates on the rate of respiration in yeast. To measure the rate of respiration, an indicator called methylene blue can be used and when it accepts hydrogen ions, it changes from dark blue to colourless.

A student set up three test tubes as shown below. Each tube contains 10 cm³ yeast, 10 cm³ methylene blue and 10 cm³ of the substrate indicated.

The experiment was left for 1 hour.

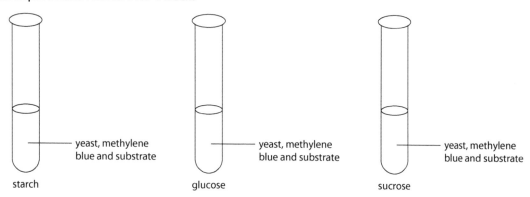

The results from each test tube are in the table below.

Respiratory substrate	Colour of methylene blue after 1 hour
starch	dark blue
glucose	colourless
sucrose	light blue

(a) Name **two** variables, not already mentioned, that should be kept constant.   **1**

_____

_____

(b) State **one** way in which the reliability of this investigation could be improved.   **1**

_____

_____

(c) Explain why the experiment was left for 1 hour before the results were recorded.   **1**

_____

_____

6. (continued)

(d) The student also added a test tube with pure water instead of a substrate, all other contents identical.

State the term used to describe this experiment and explain its purpose.

Term _____ **1**

Purpose **1**

_____

_____

(e) A similar experiment was carried out whose aim was to investigate the effect of increasing glucose concentration (%) on the time taken for methylene blue to change to colourless (seconds). The results are shown below.

Glucose concentration (%)	Time taken for methylene blue to change to colourless (seconds)
1	90
2	64
3	59
4	52
5	46
6	38
7	30

Plot a line graph to illustrate the results of the experiment. **2**

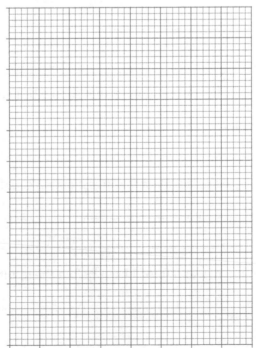

6. (continued)

(f) State **one** conclusion that can be drawn from the results of this experiment. **1**

_____

_____

7. The diagram below shows the structure of an artery.

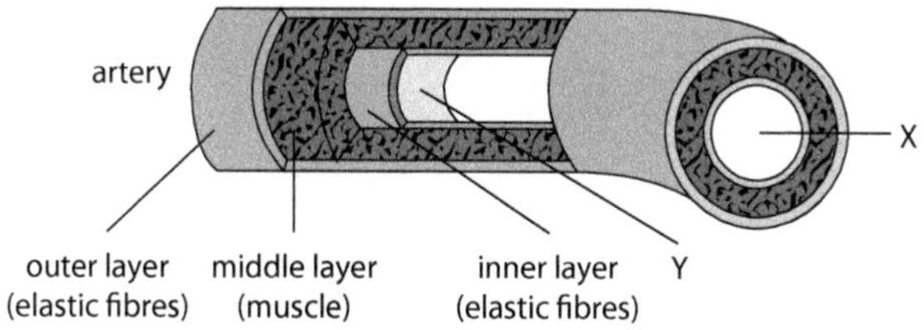

artery

outer layer   middle layer   inner layer   Y
(elastic fibres)   (muscle)   (elastic fibres)

(a) Identify the parts labelled X and Y. **2**

X _____

Y _____

(b) Give **one** function of the elastic walls of the arteries. **1**

_____

_____

(c) Changes in the smooth muscle can control blood flow through an artery.

State the term used to describe one of these changes and give an explanation of how the change affects blood flow in the vessels supplying the capillaries.

Term _____ **1**

Explanation **1**

_____

_____

(d) Give an account of the structural and functional differences between an artery and vein. **3**

_____

_____

_____

_____

_____

_____

_____

8. The diagram below shows a sphygmomanometer which measures blood pressure.

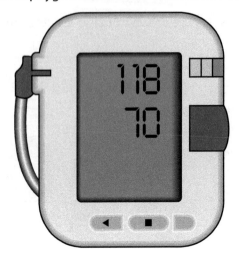

(a) State why there are two values shown on the meter. **1**

_____

_____

(b) State the typical values for a young adult. **1**

_____ mmHg

(c) High blood pressure over a prolonged period can result in hypertension.

State **one** disease in which hypertension is a risk factor. **1**

_____

9. An experiment was carried out on 20 volunteers, who were split into two groups of 10.

Group 1 were asked to count to 26.

Group 2 were asked to recite the alphabet.

Both groups were then asked to read aloud what they saw on the card below.

B

9. (continued)

The results are shown in the table below.

Group	Number of people giving each response	
	B	13
1	2	8
2	9	1

(a) Describe how the procedures used in this experiment led to the results shown. **2**

(b) Describe why the results of this experiment may not be considered reliable. **1**

10. (a) A genetic condition is caused by a change in the genome in which an extra nucleotide is inserted into the gene that codes for an enzyme.

State the general term that is used to describe a gene alteration. **1**

(b) Explain the likely effect of this extra nucleotide on the structure of the enzyme. **2**

(c) The condition occurs with a frequency of 1 in 700 males.

Assuming an equal proportion of males and females, calculate how many males are likely to have the condition in a town with a population of 350,000. **1**

Space for calculation

10. (continued)

(d) Where conditions such as this exist in a family, the family history can be used to determine the genotypes of its individual members.

Give the term used for this process. **1**

_____

(e) During in vitro fertilisation (IVF) treatment, it is possible to detect single gene disorders in fertilised eggs before they are implanted into the mother.

State the term that describes this procedure. **1**

_____

11. (a) The graph below shows the average changes in female fertility and maximum heart rate in a sample of 100 women over a period of 55 years.

The change in fertility is expressed as a percentage of the maximum potential.

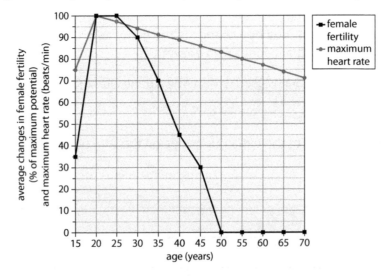

(i) State **two** conclusions which can be drawn from these results. **2**

1 _____

_____

2 _____

_____

(ii) Assuming the rate of change for the maximum heart rate stays the same, predict the maximum heart rate for an 80-year-old female. **1**

_____ beats per minute

(iii) State the two ages which show the greatest change in female fertility. **1**

_____ and _____ years

11. (continued)

(iv) State how long the females in this sample have 100% fertility. **1**

_____ years

(b) It is estimated that about 8% of women will stop menstruating by the age of 40 years. In a population of 200 women, 25% of them were found to be 40 years or older.

Calculate how many will have experienced menopause. **1**

Space for calculation

_____

12. The table below shows the number of people with type 1 or 2 diabetes in four countries of the world in 2015 and the predicted number of people with diabetes in 2040.

Country	Total population in millions (2015)	Number of people with diabetes in millions (2015)	Projected number of people with diabetes in millions (2040)
China	1,375	110	150
India	1,311	70	124
USA	320	30	35
Brazil	208	14	23

(a) Calculate the country that had the highest percentage of people with diabetes in 2015. **1**

Space for calculation

Country _____

(b) Calculate the simple whole number ratio for the number of people who have diabetes in the USA in 2015 and the projected number of people with diabetes in the USA in 2040. **1**

Space for calculation

_____ : _____

12. (continued)

(c) Tick the correct box(es) in the table below to show which statements refer to type 1 diabetes, type 2 diabetes or both. **2**

Statement	Type 1 diabetes	Type 2 diabetes
Usually occurs in childhood		
Typically develops later in life		
Occurs mainly in overweight individuals		
Individual unable to produce insulin		
Can be treated with regular doses of insulin		
Individuals produce insulin but their cells are less sensitive to it		
Blood glucose levels rise rapidly after a meal		

13. The diagram below shows some of the events associated with a cycle in reproduction.

Approximately halfway through the cycle, event A takes place.

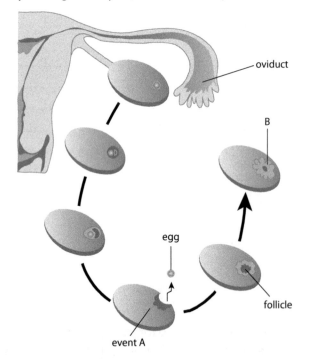

(a) (i) State the term used to describe event A. **1**

_____

(ii) Name the hormone which rises suddenly in the blood and triggers event A to take place. **1**

_____

(iii) State the gland which manufactures this hormone. **1**

_____

13. (continued)

(iv) State the name of structure B. 1

_____

(v) Name two hormones manufactured by structure B. 2

1 _____

2 _____

(b) Explain what is meant by in vitro fertilisation. 1

_____

_____

14. A student carried out an investigation to determine the memory span for numbers.

Six young people from four different age groups were each given 30 minutes to memorise a sequence of numbers and then recall as many as they could, in the correct order.

The results that they achieved are shown in the table below.

Person	Numbers recalled			
	Age 6–8	*Age 9–11*	*Age 12–14*	*Age 15–17*
1	7	9	17	27
2	9	14	15	13
3	3	12	24	9
4	6	10	8	14
5	12	19	16	10
6	5	2	22	23
Average	7		17	16

(a) Calculate the average numbers recalled by the children aged 9–11 years and write your answer **in the table** above. 1

Space for calculation

(b) State **one** conclusion that can be drawn from the results of this investigation. 1

_____

_____

14. (continued)

    (c) Describe **one** variable that would have to be kept constant to ensure a valid comparison could be made between the four groups of children.   **1**

_____

_____

    (d) Give **one** possible reason why children in the age group 12–14 years had a higher average than children in the age group 15–17 years.   **1**

_____

_____

    (e) Describe **one** way in which the student could adapt her investigation to ensure that these results are reliable.   **1**

_____

_____

15. The table below shows the volume of blood, measured in cm^3, in a man's right ventricle at different times during one cardiac cycle.

Time (s)	Volume of blood (cm^3)
0·0	125
0·1	148
0·2	103
0·3	70
0·4	56
0·5	55
0·6	98
0·7	125

    (a) Use the data **in the table** to calculate the man's heart rate in beats per minute.   **1**

Space for calculation

_____ beats per minute

15. (continued)

(b) Use the information above to complete the table below to show whether the valves are open or closed at each of the times shown by writing 'open' or 'closed' in the appropriate boxes.   **2**

Time (s)	Valve between right atrium and right ventricle	Valve between right ventricle and pulmonary artery
0·2		
0·6		

16. Answer either A or B in the space below.

Labelled diagrams may be used where appropriate.

A   Describe hormonal control under the following headings:

(i)   onset of puberty;   **2**

(ii) control of sperm production;   **3**

(iii) control of the menstrual cycle.   **4**

OR

B   Describe immunity under the following headings:

(i)   non-specific defence mechanisms;   **5**

(ii) design of vaccine clinical trials.   **4**

Practice Paper B

Higher Human Biology

Practice Papers for SQA Exams

Practice Paper B

Human Biology
Paper 1 — Multiple choice

Duration — 40 minutes

Total marks — 25

Attempt ALL questions.
You may use a calculator.

Leckie
the education publisher
for Scotland

SECTION 1 – 25 marks

Attempt ALL questions.

1. A fragment of DNA was found to contain 10,000 nucleotide bases of which 15% are adenine.

   The percentage of guanine present in this fragment is

   A   35

   B   30

   C   70

   D   15

2. Which of the following shows the correct structure of RNA?

	Bases	Strand(s)	Sugar
A	ATGC	single	deoxyribose
B	AUGC	double	deoxyribose
C	ATGC	double	ribose
D	AUGC	single	ribose

3. Proteins are formed by chains of amino acids.

   Name the type of bond that joins amino acids into a chain.

   A   Peptide

   B   Hydrogen

   C   Sugar–phosphate

   D   Disulphide.

4. In one experiment, amplification of DNA by PCR commenced with 500 DNA molecules in the reaction tube.

   How many DNA molecules were present after five cycles of PCR?

   A   1,600

   B   2,500

   C   8,000

   D   16,000

5. Which of the following does **not** describe catabolism?

A   Releases energy

B   Involves the breakdown of molecules

C   Has reversible and irreversible steps and alternative routes

D   Involves biosynthetic processes.

6. The diagram below represents a mitochondrion which has been magnified 100 times.

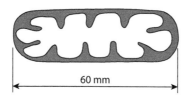

60 mm

What is the actual length of this mitochondrion? (1 mm = 1000 micrometres)

A   0·6 micrometres

B   6 micrometres

C   60 micrometres

D   600 micrometres.

7. Which of the numbers on the following graph represents the point at which almost all of the enzyme active sites are available for binding of substrate molecules?

A   2

B   3

C   1

D   4

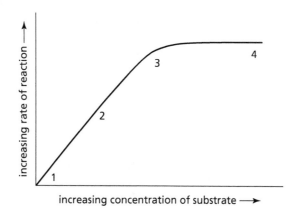

8. The diagram below shows the citric acid cycle.

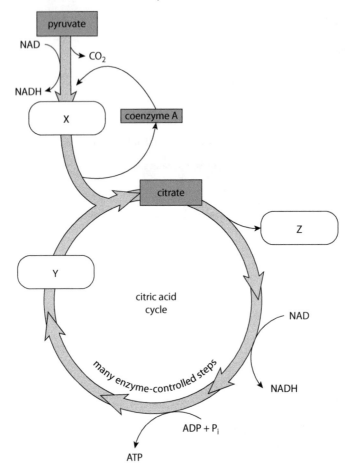

Which of the following correctly identifies X, Y and Z?

	X	Y	Z
A	acetyl coenzyme A	oxaloacetate	hydrogen
B	glucose	acetyl coenzyme A	oxaloacetate
C	acetyl coenzyme A	oxaloacetate	carbon dioxide
D	glucose	carbon dioxide	oxaloacetate

9. During the citric acid cycle dehydrogenases catalyse

A   the removal of hydrogen ions from $NADH_2$

B   the transfer of hydrogen ions to the electron transport chain

C   the removal of hydrogen ions and electrons from glucose

D   the removal of hydrogen ions and electrons from a substrate.

10. Which of the following correctly matches the reproductive organ, tissue or cell with its correct function in the reproductive tract?

	Organ/Tissue/Cell	Function
A	seminiferous tubules	produce testosterone
B	prostate gland	produce sperm
C	interstitial cells	secrete luteinising hormone (LH)
D	seminal vesicles	secrete fluids that maintain the mobility and viability of the sperm

11. Which of the following treatments for infertility involves the head of the sperm being drawn into a needle and injected directly into the egg to achieve fertilisation?

   A   Artificial insemination

   B   Intracytoplasmic sperm injection (ICSI)

   C   In vitro fertilisation (IVF)

   D   Stimulating ovulation.

12. Which of the following antenatal tests can produce cells for culturing to produce a karyotype for the diagnosis of a range of conditions?

   A   Amniocentesis

   B   Rhesus antibody testing

   C   Biochemical tests

   D   Ultrasound imaging.

13. The diagram below shows the heart during three phases of the cardiac cycle.

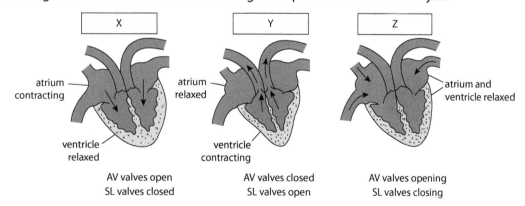

Which of the following correctly identifies phases X, Y and Z?

	X	Y	Z
A	atrial systole	diastole	ventricular systole
B	ventricular systole	atrial systole	diastole
C	atrial systole	ventricular systole	diastole
D	diastole	ventricular systole	atrial systole

14. Which of the following terms describes a clot which breaks loose and travels through the bloodstream until it blocks a blood vessel?

A   Thrombus

B   Fibrin

C   Embolus

D   Atheroma.

15. The diagram below represents a cross-section of the human heart.

The correct position of the sino-atrial node is

A   1

B   3

C   4

D   2

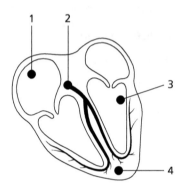

16. Scientists conducted an investigation into the effect of drinking different amounts of alcohol on the risk of developing heart disease.

The graph shows the results of the investigation.

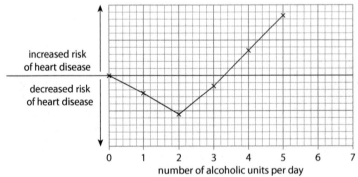

How many alcoholic drinks per week would result in the greatest chance of heart disease?

A   2

B   5

C   6

D   35

17. The changes in the blood pressure (mmHg) of a patient over time (in seconds) can be seen in the graph below.

The heart rate of this patient is

A   124 bpm

B   72 bpm

C   60 bpm

D   75 bpm.

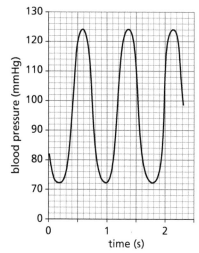

18. Doctors investigated the effect of taking glucose by mouth compared with injecting the same mass of glucose directly into the blood of type 2 diabetic people and non-diabetic people.

The doctors:

- investigated type 2 diabetic people and non-diabetic people

- monitored the concentration of insulin in the blood for three hours following the intake of glucose.

Using both type 2 diabetic people and non-diabetic people in this study

A   allowed a valid comparison

B   increased reliability

C   increased accuracy

D   ensured the variables were kept constant.

19. In order to produce a karyotype, samples taken from a pregnant woman can be

A   obtained from a scan at 8 to 14 weeks

B   amplified to obtain sufficient cells

C   cultured to obtain sufficient cells

D   obtained from a scan at 18 to 20 weeks.

20. The diagram below shows a motor neuron.

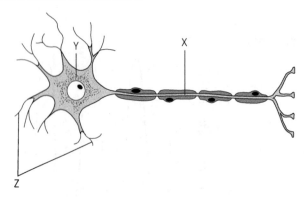

Which of the following correctly identifies structures X, Y and Z?

	X	Y	Z
A	cell body	dendrite	axon
B	axon	dendrite	cell body
C	dendrite	axon	cell body
D	axon	cell body	dendrite

21. The following diagram represents seven neurons in a neural pathway.

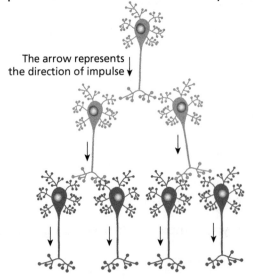

The arrow represents the direction of impulse ↓

Which line in the table correctly describes this pathway?

	Type of pathway	
A	motor	divergent
B	sensory	convergent
C	reverberating	divergent
D	motor	convergent

22. The diagram below shows an immune response.

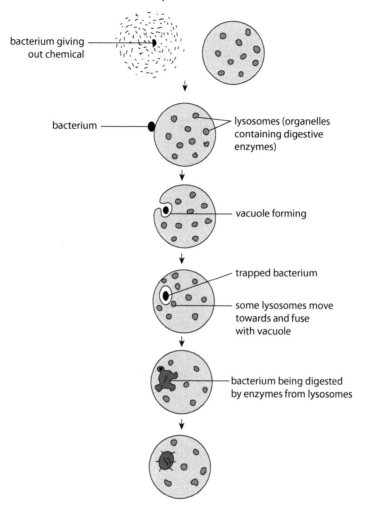

Name the type of cell which carries out this immune response.

A   Phagocyte

B   Mast cell

C   T lymphocyte

D   B lymphocyte.

23. Which of the following design features would be used to eliminate bias from clinical trials?

A   Use of placebo

B   Randomisation

C   Large sample size

D   Replicates.

24. The table below contains information about the number of cases of tuberculosis (TB) in a city over five years.

Year	TB cases in June	TB cases in December
2009	220	780
2010	445	620
2011	450	1200
2012	135	125
2013	150	400

Which of the following conclusions can be drawn from the data in the table?

A   There are always more cases of TB in December than in June

B   The number of cases of TB increased by 60% between June and December of 2010

C   The greatest percentage decrease in TB cases occurred between June of 2010 and December of 2010

D   The greatest percentage increase in TB cases occurred between June 2009 and December 2009.

25. Whooping cough is a disease that affects some infants. Doctors collected data relating to whooping cough between 1965 and 1995.

They collected data for:

- the number of cases of whooping cough reported

- the percentage of infants vaccinated against whooping cough.

The graph shows the data collected by the doctors.

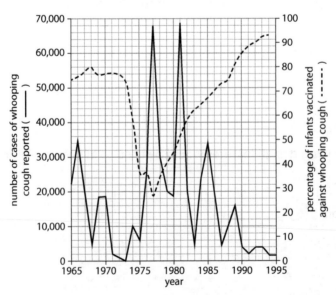

Which of the following best describes the general relationship between the number of cases of whooping cough reported and the percentage of infants vaccinated against whooping cough?

When the percentage of infants vaccinated is above

A   50% the number of cases is always below 40,000

B   50% the number of cases is always above 40,000

C   60% the number of cases is always above 40,000

D   60% the number of cases is always below 40,000.

Higher Human Biology

Practice Papers for SQA Exams

Practice Paper B

**Human Biology
Paper 2**

Duration — 2 hours and 20 minutes

Total marks — 95

Attempt ALL questions.
You may use a calculator.
Question 14 contains a choice.

Leckie
the education publisher
for Scotland

PAPER 2 – 95 marks

Attempt ALL questions.

1. The diagram below shows two strands of DNA.

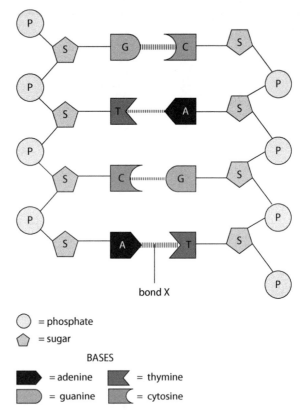

```
○  = phosphate
⬠  = sugar
        BASES
▰  = adenine      ◄  = thymine
▭  = guanine      ◖  = cytosine
```

(a) Name the type of bond shown at X.   1

_____

(b) State the term used to describe the unit made up of a sugar, phosphate and a base.   1

_____

(c) State the name of the sugar found in a strand of DNA.   1

_____

(d) Describe the directional arrangement of the two strands of DNA.   2

_____

_____

(e) State the role of DNA polymerase in the process of DNA replication.   1

_____

_____

1. (continued)

(f) State the name of two molecules, other than DNA polymerase and the template DNA, which are essential for DNA replication. **2**

_____

_____

(g) Explain the importance of DNA replication in cells. **1**

_____

2. Gene expression involves the transcription of the genetic code on a DNA molecule to form a complementary mRNA sequence.

(a) Transcribe the DNA sequence below into its primary RNA transcript by writing the complementary mRNA sequence in the table. **1**

	Triplet 1	Triplet 2	Triplet 3	Triplet 4	Triplet 5	Triplet 6	Triplet 7	Triplet 8	Triplet 9
DNA	TAC	GGA	TGC	AGA	ACG	GAC	AGG	AAG	ATT
Primary mRNA transcript									

(b) A single gene mutation occurred which added a base to triplet 2.

(i) State the name given to this type of mutation. _____ **1**

(ii) Explain the effect on the structure of the protein produced from this altered DNA sequence. **1**

_____

(c) Triplets 1, 5, 6 and 8 are coding regions.

State the general term for a coding region. **1**

_____

(d) Knowing that 1, 5, 6 and 8 are coding regions, write the sequence of the mature mRNA transcript below. **1**

_____

(e) Nucleotide substitution mutations can also occur at splice-sites.

Describe **two** effects of a splice-site mutation on the mature mRNA transcript. **2**

Effect 1 _____

Effect 2 _____

3. Below is a sequence of bases from a volunteer's DNA who had her entire genome sequenced.

TAGTTGACCATGCATGTTCAGAGC

(a) Bioinformatics can be used to compare the volunteer's genome sequence with one which is known to increase the risk of developing a particular disease.

Suggest why this may be useful in this case. **1**

_____

_____

(b) State what resources would be needed in order to identify the genes to which the volunteer's sequence of bases, similar to that shown above, belongs. **1**

_____

(c) If a positive match was found for such a volunteer who then developed the disease, in the future personalised medicine could be used.

State an advantage that personalised medicine may have over existing medicine. **1**

_____

_____

(d) State the term for the process of deciding which drug to use to treat a disease which might be linked to the volunteer's genome sequence. **1**

_____

4. Stages in the development of cells that have specialised functions are shown below.

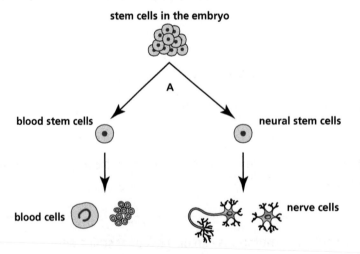

4. (continued)

(a) Identify the process that occurs at point A. _____ **1**

(b) Explain how gene activity leads to specialised functions. **2**

_____

_____

(c) State the general term used to describe cells which can become all cell types. **1**

_____

(d) A leading drug company has developed a drug which may help to treat a rare disease of the blood. They will need to carry out clinical trials to see how effective the drug is before seeking approval to make it commercially available. They have decided to test their drug on embryonic stem cells.

Suggest an ethical reason why this decision may not be acceptable to some people. **1**

_____

5. During strenuous muscular activity, the cell rapidly breaks down its reserves of ATP to release energy. Muscle cells have an additional source of energy which can be used to replenish ATP pools during strenuous bouts of exercise.

(a) State the type of metabolic reaction that occurs in muscle cells. **1**

_____

(b) During strenuous exercise, the muscle cells do not get sufficient oxygen to support the electron transport chain. Under these conditions, describe the conversion that occurs within the muscle cells. **1**

_____

_____

(c) Explain what is meant by the term 'oxygen debt'. **2**

_____

_____

5. (continued)

(d) Samples of muscle cells from two different athletes were taken and are shown below.

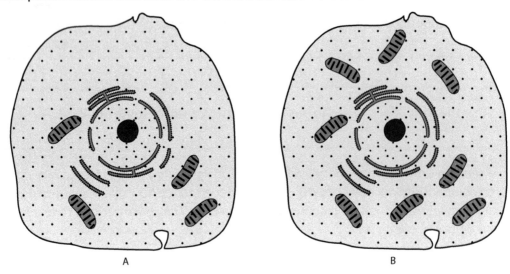

State which of the two cells belongs to an endurance athlete and give a reason for your selection.

Cell _____ **1**

Reason _____ **1**

_____

(e) There are two types of muscle fibres.

Name both and describe **two** differences between them.

Types of muscle fibres _____ **1**

_____

Differences 1. _____ **2**

_____

2. _____

_____

6. Obesity is a major risk factor for cardiovascular disease and type 2 diabetes. The table below shows the incidence rate of type 2 diabetes (per 100,000 population per year) in Scotland between 2007 and 2014 in different age ranges (in years).

Age range (years)	2007	2008	2009	2010	2011	2012	2013	2014
<10	0	0	0	0	0	0	0	0
10–19	4	3	4	3	2	3	1	3
20–29	23	24	24	24	23	29	25	27
30–39	101	127	120	116	113	146	124	115
40–49	304	310	335	321	322	347	323	315
50–59	608	628	653	629	595	650	606	570
60–69	955	917	944	851	814	833	783	733
≥ 70	835	874	867	799	747	729	762	649
Average	354	360	368	343	327	342	328	302

(a) Describe the relationship between age and incidence of type 2 diabetes in 2014. **2**

(b) Give **one** possible reason for the relationship between age and incidence of type 2 diabetes, as described in (a). **1**

(c) Calculate the percentage change in type 2 diabetes incidence for those aged 30–39 between 2007 and 2010. **1**

Space for calculation

_____%

6. (continued)

(d) Predict the incidence of type 2 diabetes for those aged 60–69 in 2015.　　　**1**

Space for calculation

_____ per 100,000 population per year

(e) Calculate the simple whole number ratio for the incidence of type 2 diabetes in 2009 for the age range of 10–19 years and 40–49 years.　　　**1**

Space for calculation

_____ : _____

(f) Identify the range of values recorded for incidence of type 2 diabetes for the age range 50–59 years over the time period 2007–2014 shown.　　　**1**

_____

7. Many couples can conceive a child within a few menstrual cycles but some couples are unable to. This could be due to a fertility problem with either the male or female reproductive systems.

In vitro fertilisation (IVF) is a widely used treatment for infertility.

At one IVF clinic, over 1,000 treatment cycles were monitored. The percentage of live births per treatment cycle was recorded.

The results were recorded against the age of the women and are shown below.

Age of women (years)	Percentage of live births per treatment cycle
Under 34	27·6
34–36	22·3
37–39	18·3
40–42	10·0
Over 42	5·0

(a) Describe the general trend shown in this table.　　　**1**

_____

_____

7. (continued)

(b) Calculate the percentage decrease in percentage of live births per treatment cycle between women under 34 years compared with 40 to 42 years.   **1**

Space for calculation

_____%

(c) Describe **one** reason that couples may experience difficulties conceiving due to infertility.   **1**

_____

_____

(d) Aside from IVF, describe **one** other fertility treatment that couples may use.   **1**

_____

(e) Couples who do not wish to conceive a child may use contraception.

State **one** example of a physical method of contraception and **one** example of a chemical method of contraception.

physical _____   **1**

chemical _____   **1**

8. Each year, between 1980 and 2000, health workers in an African country recorded:

• the percentage of the population infected with the human immunodeficiency virus (HIV)

• the number of new cases of tuberculosis (TB) per 100,000 people per year.

The results are shown below.

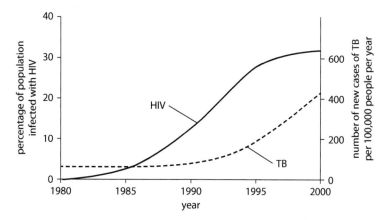

8. (continued)

(a) The number of new cases of tuberculosis (TB) per year was recorded as number per 100,000 of the population.

Give **one** reason why.   **1**

_____

_____

(b) Give an account of how HIV affects the immune system.   **3**

_____

_____

_____

_____

9. 10 students took part in an experiment where they were asked to complete 10 Sudoku puzzles. They completed the puzzles in a room alone.

Five of the students were rewarded after they completed each individual puzzle with a verbal announcement of 'well done' from a speaker. The other five received no such reward.

The time taken in seconds for each student to complete each Sudoku puzzle was recorded.

The results of the experiment are shown below.

Rewarded

Student	Time to complete Sudoku puzzle (seconds)									
	1	2	3	4	5	6	7	8	9	10
1	305	270	280	253	207	202	185	190	187	190
2	506	498	456	434	423	410	407	402	410	415
3	435	406	340	395	376	362	290	307	298	250
4	620	630	606	580	567	550	520	404	397	456
5	107	140	132	116	114	127	143	103	097	109
Average	395	389	363	356	337	330	309	281	278	284

9. (continued)

Not rewarded

Student	Time to complete Sudoku puzzle (seconds)									
	1	2	3	4	5	6	7	8	9	10
6	256	243	232	210	214	230	237	238	240	247
7	403	385	356	345	354	340	370	362	389	392
8	506	482	493	453	467	432	408	427	432	456
9	197	180	156	145	134	130	156	145	125	121
10	150	145	134	187	123	134	120	145	132	147
Average	302	287	274	268	258	253	258	263	264	273

(a) Calculate which group, rewarded or not rewarded, showed the greatest percentage decrease in the time taken to complete the Sudoku puzzle. **1**

Space for calculation

Group _____

(b) State the student with the lowest individual time across the 10 attempts. **1**

_____

(c) Describe the overall trend for both groups. **1**

_____

_____

(d) Name **one** variable that should be controlled when setting up this experiment. **1**

_____

(e) Name **one** variable that should be kept constant during this experiment. **1**

_____

(f) State **one** way in which the reliability of the results from this investigation could be improved. **1**

_____

_____

9. (continued)

(g) Draw a bar graph to illustrate the average results of the rewarded experiment.　　**2**

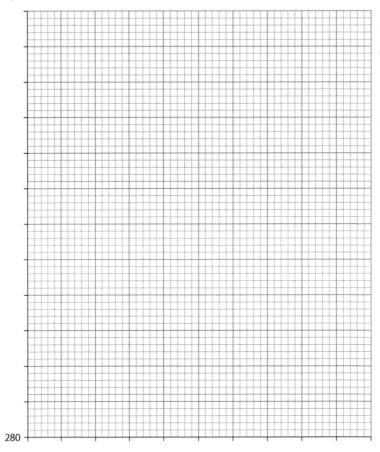

280

(h) State **one** conclusion that can be drawn from the results of this experiment.　　**1**

_____

_____

10. Scientists investigated the effect of a 6-week exercise programme on the resting heart rate of males and females.

The scientists recruited a large group of male volunteers and a large group of female volunteers. They measured the resting heart rate, in beats per minute, of each volunteer before the exercise programme. Both groups took part in the same exercise programme.

The scientists measured the resting heart rate of each volunteer after the exercise programme.

The scientists determined the mean resting heart rate and the range of resting heart rates for each group before and after the exercise programme.

10. (continued)

The graph below shows their results.

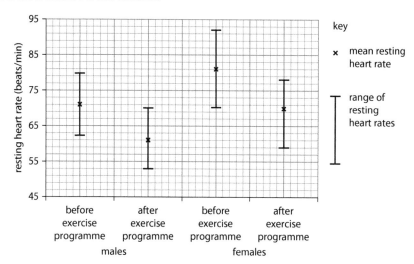

(a) Give the range of the resting heart rates in males after the exercise programme in beats per minute. **1**

_____ beats per minute

(b) Calculate the percentage decrease in the mean resting heart rate of females after the exercise programme. **1**

Space for calculation

_____%

(c) The scientists used the percentage change in the mean resting heart rate after the exercise programme to compare the results for males and females.

Explain why they used percentage change in the mean resting heart rate. **1**

_____

_____

(d) The scientists calculated the cardiac output of the volunteers before and after the exercise programme. In some volunteers, their cardiac output stayed the same, even though their resting heart rate decreased.

Explain how their cardiac output could stay the same even when their resting heart rate had decreased. **1**

_____

_____

11. 40 students took part in an investigation to test their short-term memory.

They were asked to listen to a digital pre-recording of a series of random letters (i.e. that did not spell a word). They then had to write down the letters that they heard, but only once the sequence was complete. Their scores were grouped together and recorded below.

Number of letters in sequence read out	Number of students who recalled all the letters in the correct order
1	40
2	40
3	39
4	40
5	32
6	27
7	19
8	7
9	5
10	3

(a) Draw a bar graph to illustrate the results of the experiment.　　　　**2**

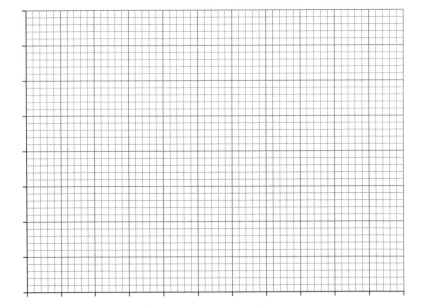

(b) Predict the number of students who would correctly recall a sequence with 11 letters in it.　　　　**1**

_____

11. (continued)

(c) Using data **from the graph**, describe the relationship between the number of letters in the sequence and the number of students who recalled all the letters in the correct order. **1**

_____

_____

(d) Calculate, as a simple whole number ratio, the number of students who recalled all the letters in the correct order when four letters were in the sequence compared to when five letters were in the sequence. **1**

Space for calculation

_____ : _____

(e) Explain why a pre-recording was used rather than someone reading aloud live. **1**

_____

_____

(f) Explain why 40 students were used and their results grouped together. **1**

_____

_____

(g) State **one** method which could be used to improve short-term memory. **1**

_____

(h) State the term used to describe the loss of items from short-term memory. **1**

_____

12. 1,000 volunteers took part in a clinical trial for Alzheimer's disease.

500 volunteers were given a trial medication for six months while the other 500 were given a tablet which did not contain the trial medication.

A computer program allocated the volunteers to one group or another and neither the volunteers nor the doctors knew which group they were in.

(a) State the name given to the tablet which did not contain the trial medication. **1**

_____

12. (continued)

   (b) Explain the purpose of giving this tablet to half the group.   **1**

_____

_____

   (c) State the term used to describe a trial where neither the doctors nor the volunteers know which group they are allocated to.   **1**

_____

   (d) Explain why a computer program, rather than the doctors, was used to allocate the volunteers.   **1**

_____

_____

   (e) Explain why a large group size was chosen for this trial.   **1**

_____

_____

   (f) State the design feature in a clinical trial which reduces bias in relation to gender and age.   **1**

_____

13. A range of white blood cells constantly circulates, monitoring the tissues. If tissues become damaged or invaded, cells release cytokines which increase blood flow, resulting in specific white blood cells accumulating at the site of infection or tissue damage.

   (a) State the name of the cytokine releasing cell.   **1**

_____

   (b) Describe the main events that occur in clonal selection.   **2**

_____

_____

_____

   (c) Type 1 diabetes is an autoimmune disease in which T lymphocytes attack insulin-producing cells in the pancreas.

      Describe why this happens.   **2**

_____

_____

13. (continued)

(d) Many people suffer from allergies such as hay fever.

Explain what happens in an allergic response. **2**

_____

_____

_____

14. Answer either A or B in the space below.

Labelled diagrams may be used where appropriate.

A   Discuss nerve transmission under the following headings:

    (i)   neurotransmitters at synapses; **4**

    (ii) neurotransmitters, mood and behaviour. **5**

OR

B   Describe heart structure and function under the following headings:

    (i)   the cardiac cycle; **5**

    (ii) cardiac conducting system. **4**